Unarmed into Battle
Story of Air Observation 1794-1986

Unarmed into Battle
Story of Air Observation 1794-1986

Atma Singh

Centre for Land Warfare Studies
New Delhi

KW Publishers Pvt Ltd
New Delhi

The Centre for Land Warfare Studies (CLAWS), New Delhi, is an autonomous think tank dealing with contemporary issues of national security and conceptual aspects of land warfare, including conventional and sub-conventional conflicts and terrorism. CLAWS conducts research that is futuristic in outlook and policy-oriented in approach.

Centre for Land Warfare Studies
RPSO Complex, Parade Road, Delhi Cantt, New Delhi 110010
Tel: +91.11.25691308 **Fax**: +91.11.25692347
Email: landwarfare@gmail.com **Website**: www.claws.in

Copyright © 2012, Centre for Land Warfare Studies.

All rights reserved. No part of this publication may be reproduced, stored in a retrieval system, or transmitted in any form or by any means, electronic, mechanical, photocopying, recording or otherwise, without the prior written permission of the copyright owner.

The views expressed are the personal views of the authors and do not have any official endorsement. They do not necessarily represent the views of CLAWS.

The author readily acknowledges the copyright of those whose works, published or unpublished, have been quoted in this book.

ISBN 978-93-80502-86-1

Contents

	Foreword	ix
	Preface	xi
1.	Balloons to World War I	1
2.	Between the Two Great Wars	25
3.	World War II	35
4.	Air Observers in the US Army	81
5.	Air Observation in the Indian Army	101
6.	Postscript	137
	References	141

Major General Jack Parham

*This book is dedicated
to
Major General H J Parham CB, CBE, DSO, British Army
whose book 'UNARMED INTO BATTLE' inspired me
and
all the Unarmed Air Observers who have dared to
go to war in the unarmed aerial platforms since
the balloon era.*

Foreword

I am privileged to have first had experience of the excellent work done by No. 1 (Independant) Air OP Flight under the command of Major H S Butalia (Butch) during the 1947-48 J&K Operations where I was a troop commander in 11 Field Regiment deployed in Naushera sector in support of 50 Para Brigade. During January 1948, one day, Butch landed at the Naushera airstrip and agreed to fly me to look at the enemy positions in front of us from his Auster MK V aircraft. Butch was kind enough to let me register Kalsian village, believed to be the Headquarter of the raiders in the Naushehra-Jhangar valley. The landing of shells in the area caused a stir and it was cheering to see the raiders running behind the big boulders and into the dry nalah.

During the 1971 war, the Air OP was equipped with the fixed wing aircraft and helicopters. During the 1971 operations in Bangladesh, I was GOC 20 Mountain Division. I made extensive use of the Air OP helicopter located at Balurghat near my divisional headquarter to reach out to the forward troops. It was fascinating to see enemy tanks and troops which indicated to me the line of the dispositions of the enemy troops.

The author has traced the story of air observation and reconnaissance from 1794 to the Second World War in detail and also briefly covered the part played by the Air OP pilots in the wars fought in our sub-continent since independence, till the formation of the Army Aviation Corps on 1 November 1986. He gives the actual battle accounts of the air observers who went up in the unarmoured platforms, first the balloons and then rudimentary flying machines in the First World War. The art of air observation and direction of artillery fire was subsequently perfected during Second World War when army pilots flew improved flying machines. Actual battle accounts of the Air OP pilots particularly during Second World War and the wars fought in this sub-continent make a fascinating reading.

The author had been sharing his research, the role and employment of the Air OP in our context with me for the last 4 years. I suggested to him to write a book to share his valuable research because most of these facts were not known to the military readers. I am glad that he took note of my advice and has come out with this fascinating story during the Silver Jubilee year of the Army Aviation Corps. I feel that this story is not only about 'Air Observation' but about the evolution of military aviation during the last 200 years and must be read by all aviators and servicemen.

Major General **Lachhman Singh Lehl** PVSM, VrC (Retd)

PREFACE

I came to Air OP (Observation Post) rather late, after exactly seven years of regimental service but had a very long inning of 18 years in Air OP and Army Aviation.

During the Air OP conversion course in 1962, our Chief Flying Instructor, Maj (later Brig) HS Sihota told us, "Gen JN Chaudhary (then Chief of the Army Staff), is very keen on the Army Aviation Corps, it should come through soon, your future lies in Army Aviation." He also made us read a book, *Unarmed into Battle: The Story of the Air Observation Post* written by Maj Gen H J Parham CB, CBE, DSO and EMG Belfield MA, published for the Air OP Officers' Association in the UK in 1956.

This book traces the evolution of the concept of air observation from an unarmed platform, its transition over the last 200 years and the actual battle accounts of these unarmed soldiers in the wars fought during these years.

After the course, I was posted to No. 1 (Independent) Air OP Flight at Nasik Road, where I had a very fruitful tenure of nearly three years. After some time, I looked for that book, to read it again. To my dismay, I learnt that it had disappeared from the Air OP Conversion Training Wing. I was disappointed and felt as if I had lost a treasure. But my quest for this book continued and I traced it again in the Joint Air Warfare School library in Secunderabad in 1968. When I saw that the book had not been read by anyone and nobody would perhaps read it, I told the officer-in-charge of the library that he should consider this book 'lost' and I would pay for it. He was kind enough to say that I could take it and he would write it off.

I was again posted to Nasik Road to raise No. 12 (Independent) Air OP Flight in 1970. That was the time when I really dug in my heels in Air OP. There I picked up another book from the 'Sarvatra' library: *Soldiers in the Air* by Brig Peter Mead – an Air OP pilot during World War II and the first Director, Army Aviation Corps, in the British Army. It is an autobiography, starting from his basic flying training at Jodhpur in 1943, the struggle to get the Army Air Corps and the problems faced after its formation in September 1957. It also made very fascinating reading. When I went to Deolali for the Artillery Reunion in 2010, I went to the 'Sarvatra' library and found that this book had not been read by anyone except me in 1971. I picked it up, for good, courtesy Brig Sandeepan Handa, Commandant Combat Army Aviation Training School at Nasik Road.

During my service, I had been making extensive notes on whatever I read about Air OP, Army Aviation and helicopters which came in very handy when in 1975, I got intimately involved with studies, case preparations and presentations on the creation of Army Aviation, first by Gen T N Raina, then by Gen K V Krishna Rao and Gen K Sunderjee. On retirement, I carried with me these notes with a view to compile and write something at some later stage when time permitted. After going through the resettlement blues and struggle for stability and security in the post retirement phase, which I suppose each defence officer goes through, I started digging into my archive files in 2003-04 and decided to come out with three books in the due course of time, but in 2008 I was 'stung' by a controversy on the Longewala battle fought by Air OP in the Jaisalmer sector in 1971. In response to a news item in a daily newspaper, I had simply made a statement that all the enemy tanks in that battle were destroyed by the Air Force with

PREFACE | XIII

the direction of Air OP pilots who had acted as Airborne Forward Air Controllers and there was hardly any ground battle.

Two veterans, a Lieutenant General and a Brigadier, probably 'stung' by my statement responded: "How could Air OP fly there, they would have been shot down by enemy tanks." The Brigadier even went further, "I am also a veteran, all I know is that Air OP is a sitting duck..........." Having war experience and with so much knowledge about Air OP and Army Aviation, I was really shocked to discover that we still have some 'naives' – hopefully they are only a few, who have been so ignorant about the war potential of this very important arm of the Regiment of Artillery which has performed beyond the expectations of the ground commanders in all the wars fought on this sub-continent. It was also, I felt, an insult to the army aviator fraternity as a whole and also individual pilots who have carried out seemingly impossible tasks beyond the 'call of duty' and are continuing to do so.

This is the time I changed track and decided to come out with this book, first, to educate the 'uninformed' and 'informed' and also the Army aviators of all ages. The 'uninformed' must know that Air OP and now 'Air Reconnaissance and Observation Units' of the Army Aviation Corps who still go to battle unarmed, have been, and will remain, a very potent arm in any future conventional war and the 'informed' must explore more possibilities and plan for the optimum utilisation of this versatile arm. The army aviation community should read this book as I feel that they know about themselves and their recent history, but they do not know enough about their ancestors and heritage.

However, I do have a regret. In spite of my request made in the annual Army Aviation Officers Association meeting held on 31 October 2010 and subsequent personal requests to many officers to send me their war stories, I drew a blank except Maj Gen GS Hundal. Therefore, whatever is included under "Air Observation in the Indian Army" is based on my personal knowledge and some notes left with me by the late Brig RC Butalia, my mentor, under whom I had the privilege to serve as a Flight Commander. Therefore, I am not to be blamed if the battle accounts, except for the battle of Longewala in chapter five, are too sketchy and brief and also if some other battle accounts have been left out altogether.

Lastly, this book does not claim to be the 'story' or 'history' of Air Observation Posts. As explained, there was an aim in writing it and also an aim plus – to show the stuff an 'Air OP pilot' is made of. I hope I have succeeded in this.

Maj Gen **Atma Singh**, AVSM, VrC (Retd)

1
Balloons to World War I

If you know the enemy and know yourself, you need not fear the result of a hundred battles. If you know yourself but not the enemy, for every victory gained you will also suffer a defeat. If you know neither the enemy nor yourself, you will succumb in every battle.
— Sun Tzu

I have no doubt that improved methods may be suggested and tried and that ultimately we shall find the gunners running their own balloons, looking on them first as much as integral parts of their batteries as are their own observation instruments.
— Col John Capper
Royal Engineers Commanding Officer
Air Battalion Royal Engineers 1906

French Army
The story of air observers who go into battle unarmed is not generally known even to the men in uniform but it is the glorious history of unarmed soldiers who dare to go up in unarmed machines, which goes back into the 18th century when even the steamship or steam train had not been invented. Even the historic flight of the first heavier than air aeroplane by the Wright Brothers on 17 December 1903, took place more than a century later.

On 15 October 1783, a balloon carrying a Frenchman, Jean-Francois-Pilatre de Rozier, left the soil of France[1] and became the first man to get an aerial view of the ground. Two days later, he took up with him Andre-Giroud de Villette and as a natural reaction to the use of any new invention for military purposes, even before the month was out, the passenger, de Villette, wrote a letter to the newspaper *Le Journal de Paris* recommending the balloon's suitability for military purposes: *"From that moment, I was convinced that the apparatus, at little cost, could be made useful to an Army for discovering the position of its enemy, his movements, advances and dispositions and that this could be conveyed, by a system of signals, to the troops looking after the apparatus."*

At the same time, the first British suggestion for the military employment of balloons for reconnaissance and aerial command posts, made in November 1783, and another made later by a British officer, Maj John Money, in 1790, in a formal paper, "Short Treatise on The Use of Balloons and Field Observations in Military Operations" found no takers in the War Office.

But the French who were in the forefront in the designing and manufacturing of balloons were not to lose sight of the idea given by de Villette. The "Committee of Public Safety" in Paris, taking notice of this new development, in 1793, asked the "Commission Scientifique" for advice on measures which might help France overcome her military difficulties. This was the time after the French Revolution when the French Army was in disarray, confronted by hostile Austria and Prussia.

On 14 July 1793, the "Commission Scientifique," four members of which were involved in ballooning either from the aeronautical or chemical aspect, gave its report that a number of captive balloons should be supplied to the Armies of the Republic for reconnaissance as quickly as possible

---

1. Earlier on 19 September 1783, two brothers, Joseph Michel and Etienne Jacques de Montgolfier, paper makers, had sent up in Versailles a hot air balloon with a sheep, cockerel and duck in a wicker basket.

Coutelle, a young chemist associated with the manufacture of balloons, was assigned the task of developing a military adaptation of the balloon which he demonstrated in the early spring of 1794. Coutelle was thereupon commissioned as a Captain of the artillery and asked to raise an 'Aeronaut Company'. Named "Enterprise" the balloon was launched on 10 April 1794, when Coutelle was ordered to move to Maubeuge to help the French force besieged in that town. Sortie reports of this launch stated, *"They are highly professional and full of information, largely about the enemy's artillery and working parties."* Capt Coutelle could, thus, historically, be called the first Air Observation officer.

Coutelle was then moved to support Gen Jourdan, Commander of the Army of Moselle. On 25 June 1794, Gen Morlot, a senior staff officer of Gen Jourdan observed the enemy position from the balloon for many hours. On 26 June, the Austrian garrison capitulated. The contribution of the balloon to this success was described variously: *"The balloon had an astonishing effect on the enemy"*; *"The French changed their dispositions as a result of observation from it"*; *"The balloon won the battle."* Coutelle himself said, *"in spite of the oscillation and swaying due to the wind, by use of my glasses, I was able to distinguish infantry, cavalry and artillery, their movement and, in general, their numbers."*

The "Enterprise" subsequently participated in the French enveloping movement and a river crossing south of Liege, providing quick and timely information to the ground troops. A captured enemy officer described the effectiveness of the balloon thus, *"One would have supposed that the French General's eyes were in our camp."*

About this time, orders were given to raise a second Aeronaut Company and the two companies, with increased strength, in men and balloons, were grouped together under the command of Coutelle as a Battalion Commander. In April 1795, one company joined Gen Pichagru's Army of the Rhine and Moselle and took part in the siege of Mainz in 1796 and another company took part in Gen Jourdan's advance beyond the Rhine but was captured by the Austrians at Warzburg in September 1796. After release from captivity under peace treaty terms, this company was re-equipped and in 1798, under Coutelle, sailed to join Napoleon's expedition to Egypt but Nelson sank the ship carrying the balloons and stores in Aloukir Bay. With this, the balloon era in the French Army came to an end. In 1801, the French military balloon organisation was disbanded and military interest in balloons virtually disappeared for the next 50 years.

Later, ballooning in Britain and France continued to be pursued more as a sport till 1855 when General Henry Lefroy, known to be the most scientific British soldier, proposed that a balloon be manufactured for experiments by the Royal Engineers but this also failed to interest the War Office.

American Civil War
Across the Atlantic Ocean, in the USA, balloonists had been equally active in exhibition flying and offered their services to act as observation and signal stations on the outbreak of the American Civil War. Notable amongst them were Lowe, John Wise, John La Mountain and James Alten. They all worked independently but it was Professor Thaddeus SC Lowe who got the best results from the use of balloon at the front. On 6 June 1861, he arrived in Washington DC and was invited to the White House to discuss his plans with President Abraham Lincoln. The President was receptive and promised that serious consideration would be given to the employment of balloons by the Army. One week later, Lowe conducted his first demonstration for the War Department. Attached to the White House with a line through a telegraph office and accompanied by a telegraph official and an operator, he sent the following message from an altitude of 500 feet:

Balloon Enterprise

June 18, 1861

TO THE PRESIDENT OF THE UNITED STATES

Sir:

This point of observation commands an area nearly 50 miles in diameter.[2] The city, with its girdle of encampments, presents a superb scene. I have pleasure in sending you this first dispatch ever telegraphed from an aerial station, and in acknowledging indebtedness for your encouragement for the opportunity of demonstrating the availability of the science of aeronautics in the military service of the country.

TSC Lowe

Next day, Lowe was called for a repeat performance on the White House lawn for the President and members of his Cabinet. This was the time when Confederate troops were threatening the security of the District of Columbia. Amongst the rumours and panic that Confederate troops were getting ready to invade the capital, Lowe made several ascents on 22 and 23 June but reported nothing of importance, quashing the rumours, and relaxed a tense situation by reporting the movement of the Confederates who he said were not endangering Washington.

On 2 August 1861, Lowe's programme received a boost when he received authorisation to obtain a new 25,000 cubic feet capacity balloon named "Union" which was delivered to him on 21 August. It was extensively used from 29 August to 1 October 1861 in the area around Washington. It was from this

---

2. 50 miles was an overestimation. From 500 ft, it is possible to observe an area about 15 to 20 km in diameter.

balloon that the Confederates were discovered building earth works around Washington within the range of Union guns, presenting Lowe the opportunity to adjust artillery fire from the balloon on 24 September 1861. He telegraphed his observations to an artillery officer located over three miles away. Excellent results were obtained and this was extensively practised in the subsequent campaigns.

On 25 September 1861, the War Department made the Balloon Corps part of the "Army of the Potomac" and the Secretary of War appointed Lowe as the chief aeronaut and in-charge of the new Service. He was told to construct four additional balloons. By January 1862, Lowe had seven balloons and four aeronauts. When the Confederates withdrew from the Washington area, aeronauts were sent to support the Federal troops in the other campaigns.

By now, the Confederate forces had also become aware of the effectiveness of the Federal's Balloon Corps and employed a few balloons for observation purposes but never reached the proportions of the Federal's Balloon Corps.

However, Lowe's Balloon Corps was also shortlived. In May 1863, Lowe sent observations from his balloon about the movement of Confederate troops during a battle at Chancellorsville which were either ignored or could not be used to advantage by the ground commanders, leading to friction between him and the commanders. In disgust, he left the Balloon Corps and returned to exhibition flying.

The Army ordered the Signal Corps to take over the Balloon Corps but the Signal Corps objected on the ground that it had neither sufficient funds nor the personnel to support it. As a consequence, the Balloon Corps was disbanded in June 1863. We will revert to the subsequent events in the US Army later but let us first see the events in the British Army where interest in balloons was again revived.

British Army

Capt F Beaumont of the Royal Engineers who had returned to England in 1861 after serving as a military observer with the Federal Forces in the American Civil War and his brother officer Lt George Edward Grover again revived the interest for employment of military balloons. A paper was sent to the Ordnance Select Committee which in February 1862 recommended to the War Office that trials be carried out for employment of balloons since their effectiveness as one of the resources of modern warfare was not in doubt. Typical of the British, the trials were delayed for a year as the War Office asked for the experiences of a number of other Armies in this field before giving their approval for trials.

Henry Coxwell, an experienced balloonist who had earlier made prolonged but unsuccessful efforts to persuade the War Office to use balloons in the Crimean War and elsewhere, assisted by Beaumont and Grover and with the active participation of members of the Select Committee, carried out trials at Aldershot and later at Woolwich.

In February 1865, the Ordnance Select Committee reported on the satisfactory outcome of the trials but did not recommend the special preparation of (military) balloon equipment at a time when "profound peace" prevailed on the continent of Europe.

Grover made one more effort in 1868, quoting successful balloon operations by the Brazilian Army against the Paraguayans and asked for £ 100 for the construction of a gas-generating apparatus large enough for use in the field which was also vetoed by the Secretary of State.

Till 1878, the British War Office continued to resist the proposals to buy a balloon for war purposes but then it allotted £ 150 as the cost of a balloon and engaged Capt James Templer, a militia officer and noted experienced private balloon pilot, on 10 shillings per day basis, initially to pass on his expertise to officers of the Royal Engineers, the Army's technical arm.

Capt HP Lee headed the team of Royal Engineers which also carried out vulnerability trials. After vulnerability trials in 1880, it was decided that a number of officers and men of the Royal Engineers should be trained in ballooning as aeronauts—aerial reconnaissance, photography and signalling being the part of their syllabus. By then, Templer was fully absorbed in the Army.

In 1884, an expedition was sent to Bechuanaland to deal with the Boer raiders and a balloon detachment commanded by Maj Herbert Elsdale, Royal Engineers, accompanied this force. In the following year, Templer himself took out a detachment of three balloons to Suakin to join Gen Wolseley's expedition into Sudan and it is stated that a 'six-stone' Arab, Ali Kerar, was sent up in the balloon and from a height of 2,000 ft, he reported big guns firing at Suakin.

All these years, balloon detachments operated on an ad-hoc basis without any sanctioned establishment as there were still many sceptics in the War Office and as well as in the Army who doubted their usefulness in war.

In 1886, trials were carried out on the artillery ranges at Lydd in the observation and correction of artillery fire by balloon observation which earned commendation from the Ordnance Select Committee. A proper establishment of personnel and equipment was then sanctioned for the balloon sections and by the early Nineties, ballooning detachments were fully accepted into the Army fold, with Templer attaining the rank of a Lieutenant Colonel and appointed Instructor-in-Ballooning. Later, in 1897, when the organisation had settled down in Aldershot, he became "Superintendent of the Balloon Factory."

Although the British government took almost 70-80 years to realise the potential of balloons as a military platform after the French had first used them against the Austrians in 1794, once it was convinced about the usefulness of their military employment, it went ahead unhesitatingly to develop it into military art for reconnaissance in the most professional way. Between 1890 and 1900, there was steady development in balloon training and equipment. Balloon detachments regularly attended artillery practice camps and courses were held to train officers as observers from all arms and staff. A manual of Military Ballooning was also published. Captive balloons could safely operate in winds up to 20 miles per hour at a height of about 1,000 ft with two men aboard. Communication with the ground was normally by means of reports and sketches sent down the cable in weighted bags or by telephone. The transport of one balloon section consisted of eight wagons, one for the balloon, cable and handwinch, one for miscellaneous stores and six for the hydrogen in tubes (54 tubes in all).

The British Army made extensive use of the balloons in the South African War. Three balloon sections were sent out from England to Durban and Cape Town between October 1899 and March 1900. Altogether, 30 balloons were sent. Balloon factories were set up at both Cape Town and Durban to produce hydrogen which also was supplied from England in tubes.

A balloon section under Maj GM Heath, Royal Engineers, took part in the battle outside the town of Ladysmith where part of Gen Sir George White's force was besieged. Balloons made several ascents within the besieged defences for about a month before the gas ran out, passing useful information about the enemy positions. Maj Heath, who later became a General, was awarded the Distinguished Service Order (DSO). Rear echelons of this section under Heath's second-in-command, Capt GE Phillips, which were not besieged, with the help of the Royal Navy and Royal Artillery personnel, also took part in the battle during which Philips was wounded.

Another balloon section took part in the battles of Magersfontein and Paardeberg on 11 December 1899 and 24 February 1900 where balloon observers adjusted the fire of 65 Battery (Howitzers) on the targets immediately behind the ridge (reverse slope). Although communication between the battery and balloon was in its infancy, very useful information was received from the balloon observers, including a sketch of enemy positions based on which an attack plan was made and executed, resulting in surrender by the enemy. Observer Lt AHW Grubb was awarded the DSO.

Thereafter, balloon observers were used every day to engage targets and observe enemy movements not seen from the ground. When the balloons were close to the guns, the simple expedient of shouting down observation of fall of shot with the aid of a megaphone was also tried out quite effectively. Capt GF Mac Munn, Royal Artillery, frequently took part in observation from the balloon in this operation, probably one of the first air observation posts of artillery in the British Army. Till then, the balloon observers had all been from the Royal Engineers.

On 2 May 1900, with the arrival of the 6-inch gun, two enemy camps only visible from the balloon were engaged, with observations being shouted down through a megaphone. In this operation, three well-concealed enemy guns using smokeless powder which opened up were spotted by balloon observer Capt Warry who directed the 6-inch guns, silencing the enemy guns. This performance was repeated on the following day after which there was no further enemy artillery activity. When the main divisional attack developed against the Boer's right flank, the balloon observers reported the enemy's redeployment and the movement of reinforcements.

That was the end of balloon operations in the South African War, as later, in the guerrilla warfare which followed, balloons had little opportunity to be of use.

In 1903, Lt Col John Capper, Royal Engineers, a former Secretary to the Military Aeronautics Committee, took command of the balloon sections. Proper training programmes for piloting, communications and observations were laid down. Drills were worked out so that from the time of unloading from the vehicle,

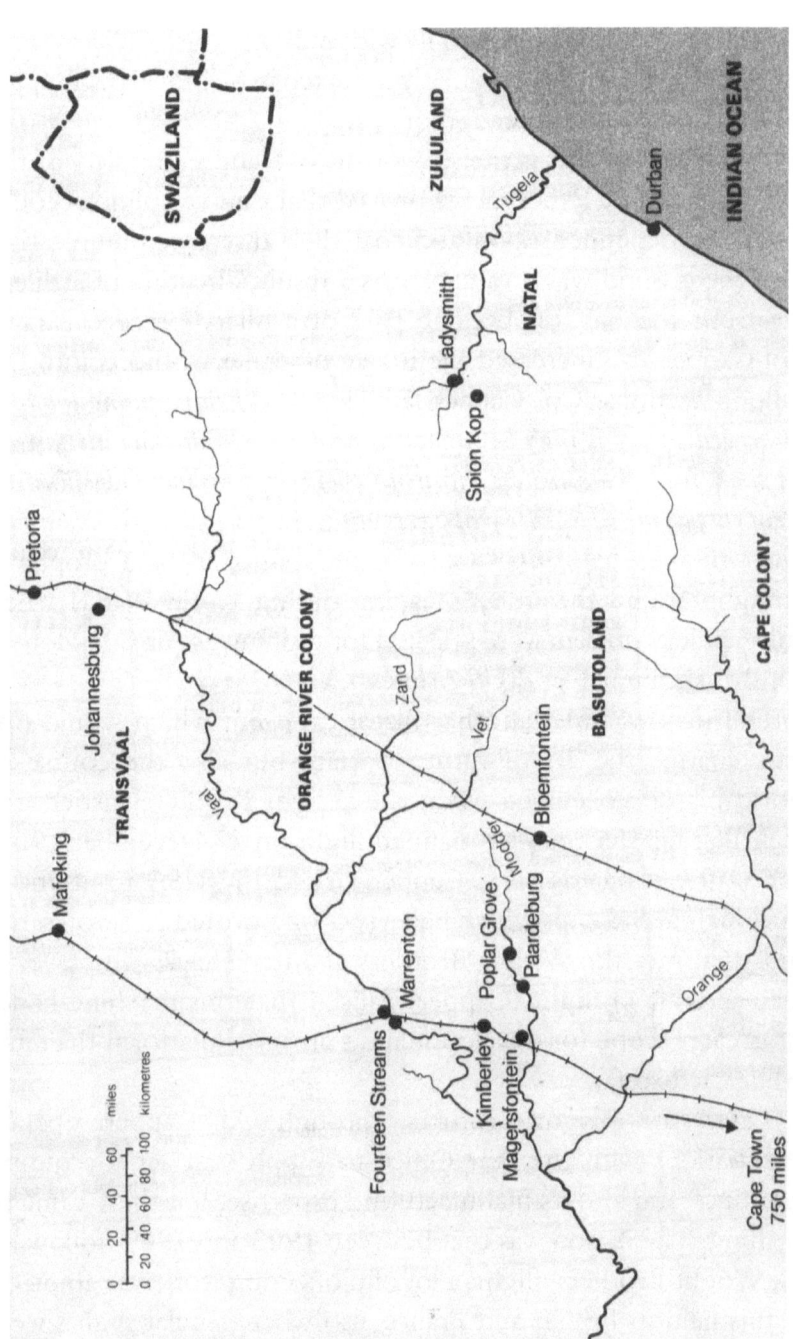

Map 1: British Army Campaigns in South Africa

the balloon could be ready to fly within 20 minutes. Officers from the Royal Artillery were regularly inducted in balloon detachments for observation and direction of fire.

A method of fire was also worked out and embodied in the training manual. Experiments proved that captive balloons could always be depended upon and that their direct communication with the ground was invaluable. As a result, directing of artillery fire from the balloon became part of artillery practice camps. Col Capper also increased the intake of artillery officers into the balloon sections. Col Capper later wrote, *"I have no doubt that improved methods may be suggested and tried and that ultimately we shall find the gunners running their own balloons, looking on them just as much as integral parts of their batteries as are their own observation instruments."* Col Capper's visionary and prophetic thoughts found their true meaning during World War II when only artillery officers were selected for training as Air OP pilots in the British, French and US Armies.

During this time, another historical event which would not only change the art of future warfare but also the course of world history, occurred when the Wright Brothers carried out their first heavier than air manned flight on 17 December 1903. Soon after, in 1904, Col Capper visited the USA to attend a world fair at St Louis which had a section devoted to aeronautics and then met the Wright Brothers at Kitty Hawk. After seeing their aircraft in flight, Capper realised that this machine had a far greater scope for development as an aerial platform than any other tried so far.

Later, the Wright Brothers, through Col Capper, opened negotiations with the War Office to supply "an aerial scouting machine" and also its manufacturing expertise, but these came to nothing. The reason was that between 1903 and 1909, following the Wright Brothers' flight, a lot of pioneering work was going on in this field in Britain and France also where similar flights were successful in monoplanes and biplanes. In the UK, Louis Bleriot made the most significant flight of 40 minutes on 24 July 1909 from Calais to Dover in his self-designed monoplane.

It was now the combination of the human endeavour, the innovative military minds and the commercial acumen of those who made these rudimentary flying contraptions which resulted in the trials of these machines for military purposes in the UK, France and USA. In 1910, during the British Army manoeuvres, amateur engineers and fliers demonstrated in their own aeroplanes, the possibilities of aeroplanes for reconnaissance, although initially there was some scepticism due to the postponement of some flights due to bad weather and because the noise of the aeroplanes could frighten the cavalry horses. At the same time, in France also, four aeroplanes took part in the corps level exercises and demonstrated their effectiveness for reconnaissance.

In May 1910, the Advisory Committee for Aeronautics was set up to advise the Admiralty and War Office on military aviation, as a result of which a Special Army Order was issued on 28 February 1911, establishing the Air Battalion Royal Engineers from 1 April in which, on Col Capper's recommendation, a company containing aeroplanes was also included. The Army by then had agreed to purchase some monoplanes and biplanes. By 1912, the battalion had 11 aeroplanes in service with a further eight under manufacture in the workshops at Farnborough. Only a few of these aeroplanes possessed reliable compasses; none had altimeters and the maximum speed of the machine was 60 miles per hour. It was also decided to induct officers from the cavalry, infantry and artillery, who showed the aptitude for aerial work, to this new outfit.

That year, major political developments were taking place in Europe and the Committee for Imperial Defence had assessed that its armed forces were unprepared for war with a major European power. Brig Gen David Henderson, Director of Military Training in the War Office, urged several remedial measures, one of which was the expansion of the Army's air reconnaissance capabilities and establishment of a central flying school for the training of pilots which were accepted by the War Office. The second was that the Air Battalion could no longer continue as a Royal Engineer unit.

Thus, in May 1912, the Air Battalion Royal Engineers was reorganised into the Royal Flying Corps (RFC) as a joint Service unit with a naval as well as a military wing but the former soon branched off as the Royal Naval Air Service (RNAS). It was also decided to raise seven companies, later to be renamed squadrons, with an establishment of 131 Army aeroplanes. An order for 26 new machines had already been placed which arrived in time for the annual manouevres in the autumn of that year.

By June 1914, out of the seven squadrons, five had been formed and put through the major training programme for the war which began in August that year.

On the outbreak of war, Brig Gen David Henderson was selected to command the "Royal Flying Corps." He underwent a flying course to qualify as a pilot at the age of 49 which was viewed as an act of 'a madman' by his friends and relatives.

Leaving one squadron for the defence of the 'homeland' against air attacks, on 13 August, four squadrons of the Royal Flying Corps (RFC) were flown from Dover to join the British Expeditionary Force (BEF) at Maubeuge in France. This maiden adventure by these flying machines had its difficulties and misadventures which included two fatal crashes before the arrival at Dover, two more crash-landings between Dover to Maubeuge, and one forced landing on the French coast due to a faulty engine where the pilot was 'captured' as a possible spy by the French Territorial Army. Even so, by 18 August, 105 officers, 755 other ranks and 45 aeroplanes were ready for the operations from Maubeuge.

The RFC had arrived just at the right time as the Field Marshal Sir John French, Commander-in-Chief (C-in-C), BEF, who had been ordered to advance into Belgium on the left flank of the French Army, had also completed the concentration on 19/20

August and he had very little intelligence about the enemy which comprised more rumours than any hard information.

It will be interesting to know that there were still voices of dissent about the viability of this new air arm. A majority among the cavalry led faction also regarded that the money spent on aeroplanes was being wasted.

World War I

The first two reconnaissance sorties by the Royal Flying Corps were launched on 19 August 1914 but the ground commanders made the mistake of launching the sorties without observers, as the pilots, with their concentration distracted due to the difficulties of aerial navigation over entirely strange country, could not bring much information about the enemy. At the same time, since the pilots were only trained to fly and not trained in the art of observation, they found it difficult to identify war-like objects on the ground. Learning lessons from these sorties, the subsequent sorties were launched on 20 and 21 August 1914 with observers on board.

Overall, twelve reconnaissance sorties were carried out each day. While the pilots concentrated on flying and navigation, the observers could identify the enemy army on the march, distinguishing among infantry, cavalry and artillery columns as well as between the headquarters and the supply units. During these and subsequent sorties, a large German Army column in an outflanking manoeuvre was spotted moving in the direction of Brussels which they entered by evening. A large body of cavalry with guns and some infantry was also reported moving south of Nivelles. A clear picture had now emerged from these reports which persuaded Sir John French to draw back and get out of a troublesome situation. It was also from these reports only that Sir John French learnt about the defeat and withdrawal of the French Fifth Army to positions well-rear of the BEF.

Although the Germans were aware that they were under aerial scrutiny and opened fire on British and French aircraft, they made no effort to conceal themselves when halted.

The performance of the observation pilots in these initial operations was commendable, which was acknowledged by Field Marshal Sir John French, C-in-C BEF, in his dispatch to the Admiralty, of 7 September 1914, *"I wish particularly to bring to your Lordship's notice the admirable work done by the Royal Flying Corps under Sir David Henderson. The skill, energy and perseverance have been beyond all praise. They have furnished me with the most complete and accurate information, which has been of incalculable value in the conduct of the operations. Fired at constantly both by friend and foe, and not hesitating to fly in every kind of weather, they have remained undaunted throughout."* This was no routine compliment.

The appearance of the aeroplane on the battlefield in the first year of the war did not make balloons irrelevant as by this time enough experience had been gained to use them as aerial military platforms. Moreover, the aeroplanes being still in their infancy, the pilots had to contend with many problems like reliability of the engines and airframes, high vibration levels which inhibited the use of telescopes and binoculars, open cockpits, susceptibility to strong gusts of winds, thermal movements of air, bad visibility and unreliable communication system.

At this stage, for proper understanding of the story of air observation, it will be appropriate to discuss briefly the type of fighting tactics followed in World War I. Offensive and counter-offensive by both Allied and German Armies, with no hope of a breakthrough by either side, led to a stalemate. By December 1914, both rivals had dug in lines of trenches which extended from the mouth of the river Yser in the North Sea to the Swiss border. These fortified trenches were protected by barbed wire, machine gun emplacements and bunkers, backed by massed artillery guns of ever-increasing calibres and range. All offensives on the

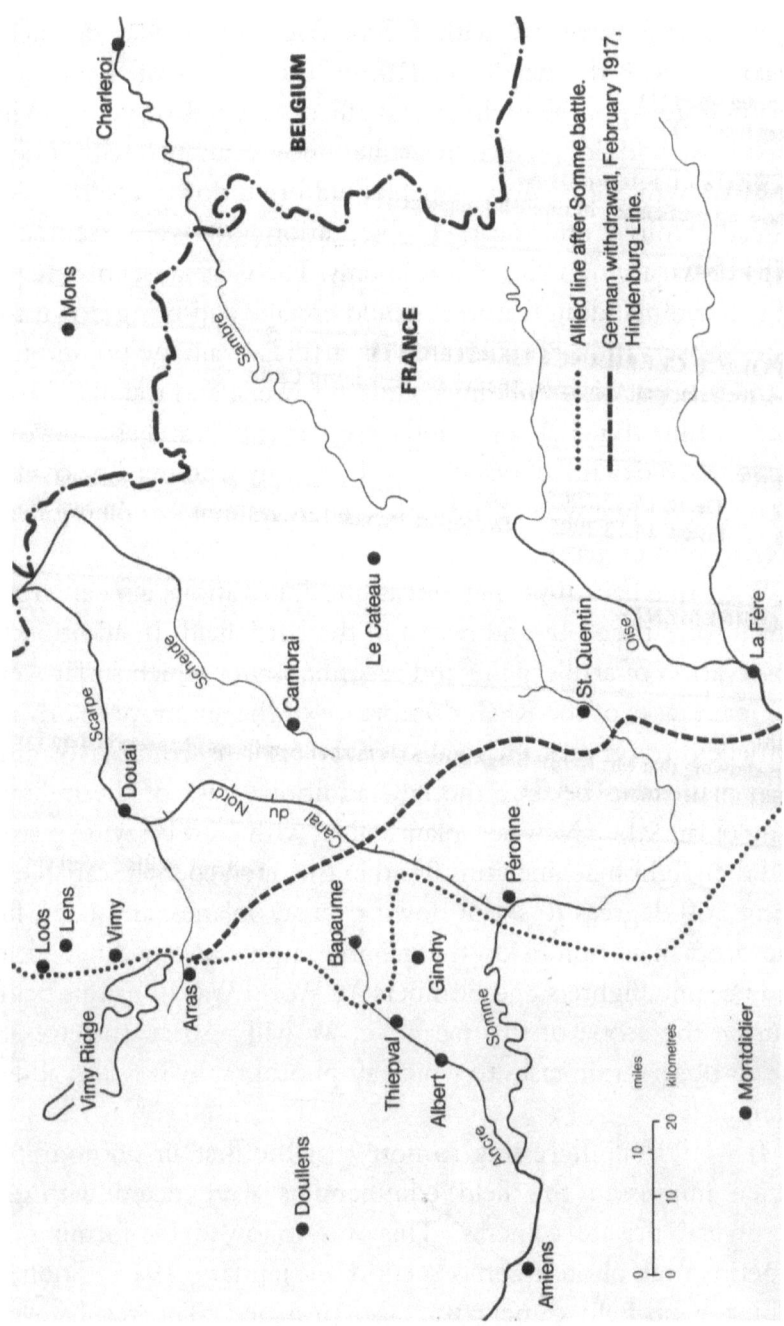

Map 2: World War I

German-held territory, with formidable fortress-like defences, failed to break the deadlock. During these years of 'siege', the machine guns, barbed wire and artillery reigned supreme. With unprecedented dominance by artillery now equipped with longer range of high calibre guns which could bring down 'indirect' fire instead of 'direct fire' as earlier, observation posts were essential to bring down accurate fire on the enemy. The Germans, entrenched on the commanding features, would engage and bring down the Allies' observation posts located on artificial vantage points such as the roofs of high buildings, church towers, and the like. From the frontline trenches, a ground observer could rarely see beyond the forward defended localities of the enemy who lived below the ground level in a web of multiple-trench systems stretching both laterally and in depth.

With this backdrop, new ideas and innovations also emerged which were tried out and tested in the battlefield. In addition to observation of artillery fire and reconnaissance which so far were the main tasks of the RFC, dominance of the air above the rival's battlefield to prevent the rival's aerial platforms from performing their main tasks became the new additional task of the military wing of the RFC. New aeroplanes, first, with crew carrying pistols and then light machine guns fitted in the open cockpit, capable of firing 360 degrees to shoot down rival aeroplanes, and later, fire and drop small bombs on the ground targets assumed new roles and became 'fighters and bombers' of World War I but this being outside the scope of this narrative, we will restrict ourselves to the air observation tasks to which air photography was also added later.

It would be interesting to note that the first air photographs which impressed the field commanders were taken with the observers' private cameras. This was followed by forming an experimental photographic section in January 1915. Soon, a robust hand-held camera was also provided. The results were beyond expectations and a photographic section was established in

Aerial platforms used by Air Observers in World War I

Aerial photograph of German Defence, 1917
taken by Army pilots of Royal Flying Corps

the RFC at the scale of one per Army. Before the famous battle of Neuve-Chapelle, the German trench system was comprehensively photographed and trench maps derived from the photographs distributed down to the sub-unit level. By the middle of 1915, a camera which could be fixed onto the aircraft and easily operated by the crew was also available. After that, there was no looking back and aerial photography as part of artillery reconnaissance was added to the main tasks of the military wing of the RFC.

Procedures were also devised so that balloons and aeroplanes could operate simultaneously and be complementary to each other. While the aeroplanes operated from landing grounds in the rear, long lines of balloons swaying at their moorings, some three or four miles behind the respective rival's trench lines, could also be seen. Whenever the weather opened up, even for a short period, the balloon observers would immediately pass the information through their wire telephony to the ground commander, whereas, at times, aeroplanes, due to bad weather on the airfields in the rear, could not even take off.

The services of one Capt Caquot, a French Civil Engineer, who had earlier performed his national service in a balloon unit in 1901, were also loaned to the RFC in 1914. He immediately designed a modified balloon with vertical fins above and below the tail for better stability which was universally adopted and used by the Allies.

In 1916, the RFC was reorganised into brigades on the basis of one brigade per Army. Each brigade had two wings, later named squadrons, one wing allotted per corps for close reconnaissance, artillery observation and photography and the second wing allotted to the field Army for deep reconnaissance, photography of areas in depth and fighter protection. Balloon squadrons, with varying numbers of balloon sections, were allotted as per the requirement of ground commanders. By now, both sides had concluded that in artillery power and its accuracy lay the best hope of achieving a break-through. On the accuracy of the artillery fire also rested the lives of the infantrymen and their success in attack. Therefore, in the trench system of warfare, artillery duels were common and became increasingly fierce during offensives and counter-offensives. Invariably, all artillery registrations and bombardments were 'observed' and directed by artillery observers from balloons and aeroplanes and the demands on the RFC for artillery spotting grew. Balloons, being more susceptible to enemy fighters, were also provided the protection of anti-aircraft guns. With the availability of long range cameras, balloon observers also produced very useful photographs of the enemy defences. During three years of the 'siege,' lots of discussions were held, problems faced, lessons learnt and procedures standardised and refined in the direction of artillery fire by the air observers, as discussed in the subsequent paragraphs.

A 'corps squadron' usually had three flights (of aeroplanes). The flights carried out impromptu shoots against fleeting targets like hostile batteries, active mortars or anti-aircraft guns and movement of transport behind the enemy's front lines. Observation of pre-arranged shoots with fire plans prepared by the counter-bombardment staff and also to support the infantry and cavalry attacks were the regular tasks of these flights. The observer was provided with a wireless transmitter on which he sent one-way signals in Morse Code to an RFC receiving set at his affiliated battery which was also overheard by another receiver at a 'central wireless station' in each Corps Headquarters (HQ). This station was also connected by telephone line to the battery to resolve communication problems between the aircraft and battery. The battery also communicated with the aircraft by means of ground-strip signals or occasionally by lamps. The corrections by the observers were sent through the 'clock-code' system. The aeroplane usually flew between 3,000 and 6,000 ft and at lower height during an unfavourable air situation. Procedures were also evolved for cooperation between aeroplanes and balloons where, at times, aeroplanes handed over the subsequent engagements or an unfinished shoot to the balloon observers.

Balloon and aeroplane pilots/observers on both sides, considering the environment, and the fragility and vulnerability of the contraptions they flew, carried out their tasks commendably. As their success became more and more apparent, the balloons became regular targets of rival fighters. Balloon crews, besides anti-aircraft guns, were provided with parachutes which, in those times, were far from infallible.

Besides the unreliable communication between the observers and the guns, the vulnerability of observers to inadvertent destruction by own or enemy shells was another problem. Slowness of shoots, particularly with aeroplane observation, was notable and the pilots were often airborne for two or three hours to engage two or three shoots.

Their casualty rate was high and so was the casualty rate of the ground troops, particularly when each side tried for a breakthrough. Nevertheless, one was left with a feeling of admiration for these pilots and observers, often young in years and experience, possessing a high degree of cold-blooded courage for taking on such responsible and daunting tasks. With force of character, intellect and high level of professionalism, they often succeeded in them.

One major lesson of World War I was that the observers with artillery experience performed their tasks more effectively. Therefore, the training pamphlets of that time emphasised the need for observers to visit the batteries to see the procedures and the problems of shooting from the gun end. Even as early as 1915, the Royal Artillery proposed to assume control of artillery cooperation flights but this proposal was not agreed to as its opponents felt that these units had to carry out many other tasks and the supply and maintenance problem must remain centralised. Another reason was that the RFC although still not completely autonomous and not the least a separate Service, felt that anything which climbed into the air must belong to them. In 1916-17 Gen Rawlinson of the Fourth Army and Gen Horne of the First Army proposed independently that 'artillery (aeroplane) squadrons' should be placed, except for purely technical matters, under the direct command of the Corps Artillery Commanders as they felt that there was a need for very high skill in the air observers, with knowledge of artillery tasks, and for intimate relations between artillery units and the air observers working in combination.

However, the fast pace of development of the aircraft industry, new weapons and their delivery systems and also to meet the new contingencies of war, with no break-through in sight, rather the German Army still having the upper hand, and with hundreds of war aeroplanes operating on both sides, dominance of the air above the theatre of war became a major objective of the military planners. All other matters were relegated to lower priority for the time being. This led to another major event – formation of an independent Service, the Royal Air Force (RAF) on 1 April 1918, by incorporating the RFC and setting up of a separate Air Ministry which would bring in revolutionary changes in the concepts of warfare in the post-war years. It is believed that one event which led quickly and directly to the creation of the Royal Air Force were the German daylight air raids on London on 13 June 1917 and 17 July 1917 in which 216 people were killed and 622 wounded. But it is also believed that this decision was based on some studies which showed that the damage inflicted on the enemy by bombing operations on targets like railways, war industries and airfields was so great that air power was recommended to be separated from the strategies of the seas and the land battles.

This was also considered by some as an emotionally inspired political decision rather than sound military judgment since the decision was taken without considering the views of the most informed and experienced men in the theatre of war in both Services – the Army and the Navy. Therefore, it was natural for the Admiralty and War Office to regain control of some elements of the new air arm in the post war years as it was proved in the war that aircraft closely integrated with naval and military functions provided the most effective support. But as proved later, with the aircraft becoming faster, more complicated and expensive, with its weapons and delivery systems becoming more devastating and also aviation technology making very fast strides, the existence of a third Service to explore the possibility of war in the third dimension was fully justified.

Thus, from 1 April 1918, all the assets of the RFC were transferred to the new independent Service – the Royal Air Force – which continued providing cooperation to the Royal Navy and the Army through the same naval and army cooperation units now merged with it. The Air Ministry, to dent the argument and soften the opposition of the other two Services, proposed to invite secondment of officers from the Royal Navy and the Army to be trained as pilots and fly in the squadrons supporting their respective Services.

2

BETWEEN THE
TWO GREAT WARS

> Unless air power is integrated with sea and land forces, in themselves Fleets and Armies lose the greater part of their fighting values.
>
> — Gen JFC Fuller (1978-1966)

British Army

The twenties and early thirties were the years of debate during which army and naval strategists started evaluating the impact of air power on the ground and naval battles. They concluded that the full scope of air power had been imperfectly realised and integration of certain elements of air power with the Army and Navy was not only essential but a great battle winning factor. "We do not want to go to war with untrained boys directing our air reconnaissance and artillery bombardments," said Gen Sir Edmund Ironside in 1924, then Commandant of the Army Staff College and later to become Chief of the Imperial General Staff in 1939. He was of the view that the field army must be complete with all essentials of war, including aircraft. Gen JFC Fuller (1878-1966) who was the first Chief General Staff Officer of the Royal Tank Corps in World War I and also became known for his military writings, further summed it up, *"Unless air power is integrated with sea and land forces, in themselves Fleets and Armies lose the greater part of their fighting values."*

While the Royal Navy won back the carrier-borne portion of its air arm in 1937, the Army was not able to push its case for its integral air arm as in the twenties, the RAF had shown full enthusiasm and an earnest desire to produce good support by its army cooperation squadrons which regularly took part in the army's brigade, divisional and command training exercises, sent detachments to artillery 'practice camps', organised lectures and demonstrations for staff college and other students.

However, later, the artillery practice camps became biennial instead of annual and because of shortage of live ammunition, army cooperation pilots found less or no allotment of ammunition for their shoots. A few other factors like the adverse effect on the career prospects of the pilots, if they stayed too long in these squadrons, the RAF developing new concepts for strategic bombing and the need for a large force of fighters for the air defence of Great Britain and conflicting opinions within the army on how to cope with the new threats, with the possibility of another war after Adolf Hitler became the German Chancellor in 1933, led to the army's requirements of an integral air arm being relegated to a lower priority and pushed into the background.

But the artillery never lost sight of the need for close air observation, a lesson firmly established in World War I and maintained a persistent demand for aerial observation manned by their own officers to direct artillery fire in the contact zone. Capt JH Parham who had served in a battery as a junior officer in the last year of World War I, writing in the *Journal of the Royal United Services Institution* in 1933 said, *"The battery commander, wishing to see a target out of sight behind a hill, is compelled to call upon a unit operating from a distant landing ground when five minutes up in the air above his own battery position even a thousand feet up would meet all his needs."*

The artillery commanders were also very clear that when they needed to send an air observer, they needed to do so at once and preferred to brief their own officer rather than a pilot of the other

Service who first had to be made available by someone in the rear and then briefed by a third person – the air liaison officer at the air base.

In November 1934, a small band of gunner officers who had learnt to fly privately at their own expense on the light aircraft also formed a Royal Artillery Flying Club which facilitated the exchange of ideas for the use of light aircraft for artillery observation and fire control. Brig HRS Massy, Brigadier Royal Artillery Southern Command, its President, Capt Charles Bazeley, its Secretary, and Maj Jack Parham, all notable gunners and pilots would discuss at great length the possibility of the gunner himself piloting the aircraft whilst also observing and communicating from the aircraft to the gun position. It was concluded that the air observer not being a gunner and being not in touch with the developing situation in the forward location could not select targets according to the tactical priorities and could not ensure that the aircraft was in a suitable position at the crucial time. This was possible only by a person who was part of the ground battle. Who should provide the aircraft, what kind of units, whether RAF, manned by gunners or purely Army Air Arm, with all the inter-Service repercussions which such a request was sure to raise, formed part of their discussions and brain-storming sessions.

Bazeley felt so strongly about this that in 1935, despite being overage, he managed by devious means to get himself seconded to the Royal Air Force to push his ideas which he never ceased to advocate and which he had already thought out in detail. His chance to put forward his ideas came when the subject for the *Annual Duncan Prize Essay for 1938-39* was announced to the Royal Artillery. The subject for the essay was: *"The battlefield of the future should present great possibilities for the use of artillery fire observed from the air, both against stationary targets such as the enemy's artillery, and moving targets such as reserves of mechanised troops. Do our present methods take full advantage of these possibilities, and if not, what steps do you consider should be taken to improve them?"*

Bazeley entered the competition, setting forth his views based on three years of thought process and discussions, with Massy and Parham advocating the case of FOP (Flying Observation Post) operated and flown by the artillery, using slow manoeuvrable, unarmed light aircraft with good take-off and landing capabilities.

In his essay, he made a remarkable case for FOP which holds good almost word for word till today, *"It is not suggested that the FOP (Flying Observation Post) should in any way replace the ground OP. Rather, it should reinforce it and make its working more efficient."*

Then he tabulated some of its advantages:

- An aeroplane (FOP) will always be available at the shortest possible notice. Therefore, air observation can be carried out immediately on coming into action.
- Concentrations of two or more troops or batteries can be called for by the gunner pilot.
- The gunner pilot will have much more exact knowledge of what is happening on his front. From his low operational height, he should be able to distinguish between friend or foe and so be able to give his infantry much better support.
- Targets can be selected according to their tactical importance.
- When condition permits, the gunner pilot will be able to fly forward over his own front line and obtain much information useful to higher formation commanders.
- The RAF would be relieved of the responsibility of providing all artillery reconnaissance and a great proportion of tactical reconnaissance over enemy territory within three or four miles of our own front line.
- When weather conditions were bad and the base of the cloud at 1,000 ft, the gunner pilot could still carry out air observation although the RAF pilot could not operate over enemy territory.

- The FOP could command the enemy's lines to a depth of three or four miles even under indifferent weather conditions and engage all targets such as the enemy's attacking infantry and tanks or his forward defended localities and targets within his defensive framework.
- The FOP would be able to move to a flank to see into hollows and valleys which are dead ground when viewed from immediately above the battery.

Although the essay was not a prize-winner, it was later published in the *Journal of the Royal Artillery*. These thoughts were quite revolutionary even to the gunners themselves.

The formal proposal for FOP was then initiated and the General Requirement was first officially stated to the War Office in 1938 by the General Officer Commanding-in-Chief Southern Command where Brig Massy was still the Brigadier Royal Artillery (BRA). He stated that *"what was needed was someone in the aircraft capable of giving executive fire orders to the guns. He must be a gunner officer. He must be in direct communication with his guns by radio telephony. He must be completely up-to-date in the tactical situation and able to take to the air in the shortest possible time. He needed, therefore, an aircraft of robust construction giving him a good all-round view, and stripped of all non-essentials so that the take-off and landing runs might be as short as possible. Its top speed might be quite low."*

The War Office put this gunner's view to the Air Ministry and asked for trials for the "Flying Observation Post." To this, the Air Ministry demurred, proposing that it would be better to seek to improve the existing RAF procedures for observation of fire, that the system then in use had stood the test of the 1914-18 War, that the existing 'clock code' system was adequate, that light aircraft could not be kept in action close to artillery units and that it was undesirable to introduce radical changes "at the present juncture" (keeping in view the prevailing international situation).

The War Office persisted in their request and in December 1938, the Air Ministry agreed to the joint trials under Air Officer Commanding 22 (Army Cooperation) Group and the Commandant, School of Artillery, Larkhill.

Bazeley and two other gunners, Capt Fielden and Capt Davenport, also seconded to the RAF, carried out the trials. Maj AK Mathews, instructor-in-gunnery, who had himself been a pilot in World War I was detailed to supervise the trials.

The Army thought the results were encouraging but the RAF were not so impressed. The War Office pressed for further trials using more suitable aircraft than the fast and heavy Audax and Lysander used in the first trials.

Further trials using the Taylorcraft, though still not really suitable but better than the fast and heavy Audax and Lysander, showed conclusively that effective fire could be reached considerably quicker than the existing procedure and that this fire could be observed up to some 8,000 yards. A trial against the Spitfire, the fastest fighter then in existence, which carried out mock attacks on gunner pilots showed that a light aeroplane, even without previous warning, had quite a good chance of dodging the fire of the modern fast fighters.

In the meanwhile, by coincidence, Massy first moved on promotion to take over as Director of Military Training at the War Office in October 1938 and then became Deputy Chief of the Imperial General Staff in October 1939 wherein he continued to press for further trials and establishment of the FOP. At the close of 1939, the War Office started formulating plans for flying training of gunner officers. The Air Ministry relented and was willing to help if further experiments were satisfactory. Thus, the first FOP unit of the RAF came into being in February 1940 and the directives setting forth its functions were drawn up by the War Office and the Air Ministry under the signatures of their respective Deputy Chiefs of Staff. The directive laid down that tests were to be made "to determine, in the light of practical experience obtained under war conditions, the possibilities and limitations of the FOP, the most suitable type of aircraft and the most suitable organisation."

The tests were to be in three stages:
- Initial training in the UK.
- Practical training in conjunction with the French Army at an Artillery Practice Camp in France.
- A final test under conditions of war to include shoots on actual German targets since the French and German Armies were already face to face in a semi-war state.

The directive also stated that the ideas and knowledge may be shared with the French Army which themselves were carrying out similar tests and, finally, the term "Air Observation Post" (Air OP) was to be used in the future instead of "Flying Observation Post" (FOP).

Air OP Experimental Flight
The experimental flight named 'D' Flight under the command of Maj HC Bazeley, after initial training in the UK, flew to the French artillery ranges at Mailly on 19 April 1940 and carried out intensive training for three weeks during which the French Army was very helpful. Bazeley was now all set for the third stage of the directive i.e. actual engagement of German targets. On 9 May 1940, along-with Maj AG Mathew, Battery Commander of the battery allotted for the experiment went for preliminary reconnaissance to the Saar front where the French and German Armies were already confronting each other. On the early morning of 10 May, the Germans launched their whirlwind offensive. This was the day when Neville Chamberlain resigned and Winston Churchill took over as Prime Minister. Mailly was also bombed on the same afternoon. The artillery designated for the trials promptly disappeared to join their formations, leaving

'D' Flight at Mailly, but Bazeley still hoped that the campaign would stabilise enough to continue the last phase of the trials. BBC news indicated that the situation was worsening, not by the day but hourly. The flight was recalled to England on 20 May 1940. The experiment had been launched too late and the case of Air OP was back to square one.

After the dilemma at Dunkirk when the Army lost all its weapons and heavy equipment and which showed the inability of the RAF to oppose the Luftwaffe effectively, with a very real threat of imminent invasion of Britain, each Service was up to its neck in the great touch-and-go struggle. The Army furiously tried to reequip itself and prepare anti-invasion and anti-tank obstacles. The RAF was busy developing radar and new fighters, bombers and new weapon systems and also totally immersed in the famous Battle of Britain.

In this trying period, 'D' Fight now located at the School of Artillery, Larkhill, and under the command of No. 22 Army Cooperation Group RAF at Farnborough, continued imparting flying training to the gunner officers. But the Air OP had its critics and sceptics who thought it was a waste of time and money to go on with it since they were convinced it could not operate under the threat of Nazi air power. The Royal Air Force also opined that the case against the Air OP was conclusive and suggested the disbandment of 'D' Flight.

It is said that placing of the right people at the right place at the right time can sometimes change the course of events and, consequently, the course of history. This was exactly what happened with the Air OP whose fate hung in the balance by a slender thread.

Air OP Resurrected

On 1 December 1940, the Army Cooperation Command came into being and 'D' Flight was placed under 70 Group of this new command. The Air Officer Commanding 70 Group, Air

Vice Marshal (AVM) TB Cole-Hamilton had seen the Air OP develop from the first experiment of 1938 and was its firm supporter, although his senior Air Marshal AS Barratt, AOC-in-C Army Cooperation Command, had other views. On the Army side, General Sir Alanbrooke (later to become a Field Marshal and Lord), not only a gunner but a master gunner and a champion of the Air OP from its earliest days, was the Commander-in-Chief Home Forces. On 28 April 1941, he wrote to the War Office that the Army needed Air OP and that 'D' Flight should be expanded forthwith to form the first Air OP Squadron. He also asked it to be made quite clear to the Air Ministry that "the Army considers them essential" and that they should be "pressed to provide the necessary aircraft at the earliest possible date." This clinched the issue and the principle was now accepted and the worst fears of Air OP supporters were set at rest as Air Marshal AS Barratt also consented but not before commenting, *"There is nothing new about this. An old horse resuscitated at War Office request. As long as they provide the bodies to be shot down, I do not mind ...."* which seemed like a note of 'demise' before the birth of a child. Mention also must be made here of Maj Gen Otto Lund (later Lt Gen Sir Otto Lund) Major General Royal Artillery Home Forces, who played an important part to get it accepted with his skilful advocacy and, subsequently, his wide knowledge of inter-service procedures helped to overcome many difficulties encountered in its formation.

After RAF resistance ceased, training of artillery pilots started on a regular basis and on 1 August 1941, 651 Air Observation Post Squadron RAF, the first operational Air OP unit was formed with 'D' Flight as its nucleus. A dual command and control system, along with responsibilities of each Service for these hybrid units, was laid out. Briefly, these units, although RAF units, came under the operational control of Royal Artillery Commanders. Their Commanding Officers and all other officers were to be Royal Artillery officers, but adjutant and equipment officers were from the RAF. The drivers and signallers with ground transport and radio sets were to be provided by the Army (Artillery) and technicians and technical support for aircraft by the RAF. This was what the Indian Army followed till 1 November 1986, when the Army Aviation Corps was formed with the takeover of all the Air OP units by the Indian Army.

3

WORLD WAR II

Anyone with experience in the forward area of battlefields knows.....where life is stripped to very real 'realities' and....ahead of all other priorities, comes the question, "what is the enemy doing, where exactly is he and where exactly are our people."

— Major General Jack Parham
Letter to the Royal USI Journal, August 1966

African Campaign

On 13 August 1942, the newly formed HQ 1 Army began the detailed planning for Operation 'Torch', the invasion of Northwest Africa. It was the first major sea-borne expedition by the Allies in which Anglo-American forces worked together under General Eisenhower, later appointed Supreme Commander Allied Forces. 651 and 654 Air OP Squadrons which appeared on the Order of Battle (ORBAT) were, to the great majority of the planners, a completely unknown entity. Confusion was further confounded as it was not clear whether they were Army or Royal Air Force units.

Maj Bazeley, first time unlucky, again got the chance to prove his concepts when he was appointed to take over 651 Air OP Squadron, the first to go into action in World War II.

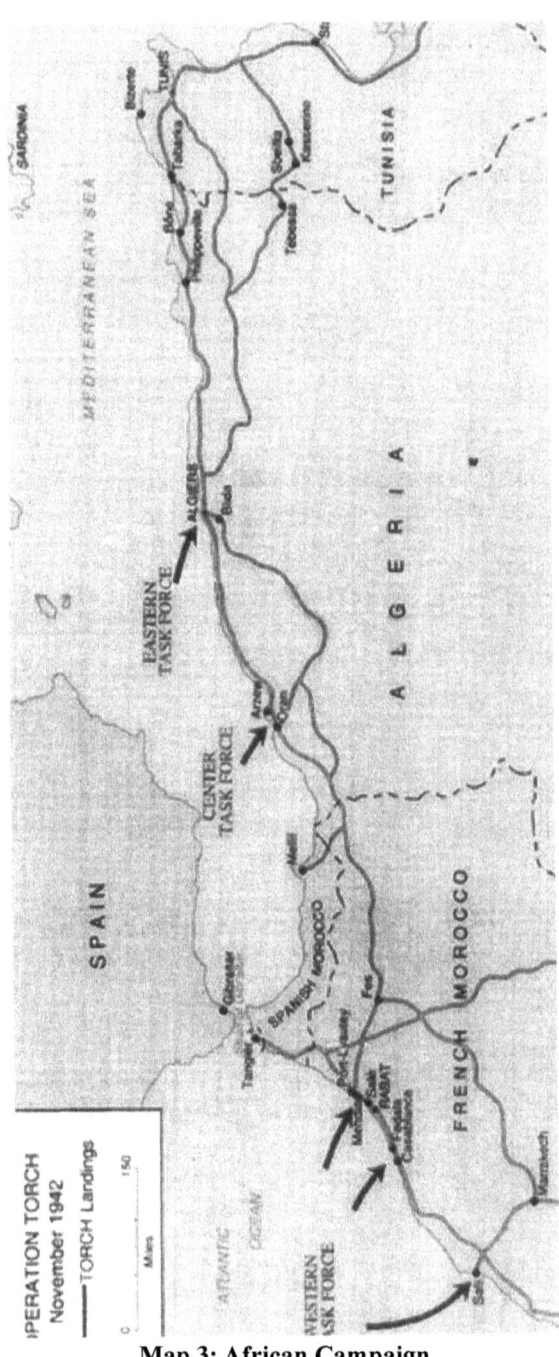

Map 3: African Campaign

After reaching Algiers on 12 November 1942, the squadron immediately went into action in support of two columns of 78 Infantry Division, making a dash for Tunis. Capt Billingham, for his daredevil actions between 28 November to 02 December 1942 was given the 'Immediate Award' of the Military Cross and Aircraftman Brown of the same squadron, a Military Medal for retrieving his vehicles and ground party between 30 November and 3 December from the Advanced Landing Ground (ALG) which had come under German fire. Thus, the Air OP won its spurs to take its place of pride as a fighting component in the very first battle.

The German attack was also repulsed between 18 and 21 January 1943 when Air OP brought down devastating fire of the complete divisional artillery, inflicting heavy losses on the Germans.

In January 1943, Maj Neathercoat took over the squadron from Maj Bazeley (later awarded the DSO) who was recalled to England as he had by far the widest knowledge of all that was involved and the Army thought it would not be right to take the risk of losing all his experience which was most needed to build up Air OP Squadrons then under raising in the UK.

It was during this campaign that certain ideas which later became the 'Standard Operating Procedure' (SOP) were tried out. The first was the 'Uncle Target', later renamed 'Uniform Target' procedure to bring down the fire of the divisional artillery, some 72 guns, within four minutes, on opportunity targets by the ground or Air OP. This was devised by Brig Jack Parham, Brigadier Royal Artillery First Army, who himself, along with Gen Massy and Maj Bazeley had mooted the idea of Air OP in 1934 while flying with the Royal Artillery Flying Club, Larkhill.

The second was to get early warning of the enemy fighters from the ground troops so that the Air OP pilot could take evasive action by using the ground terrain for protection. This was done by communicating 'Bandit' 'Bandit' 'Bandit' to the Air OP pilot followed by the indication of direction of the enemy fighters. This SOP was put to test on two consecutive days when a pilot flying at 600 ft was warned about 'Bandits' when a dozen enemy bombers, escorted by several fighters flying at about 1,500-2,000 feet approached the area. The pilot dived to ground level and flew evasively over low, dark wooded hills which he had previously chosen, suiting the aircraft's camouflage. In the second case, Capt MacGrath, while directing artillery fire, was set upon by five fighters. He took evasive action by flying to the basin of hills which had a hollow inside. He took a series of steep turns which his more highly loaded and much faster opponent could not execute. He returned to complete his task after the frustrated fighters made off.

On 4 March 1943, 654 Air OP Squadron commanded by Maj Willett landed in North Africa and after a brief period in action with the newly arrived 9 Corps of the First Army, was transferred to the Eighth Army, and by April 1943, the stage was set for the final battles to destroy Hitler's Africa Corps and their Italian allies.

By early May 1943, all was over in North Africa, and a quarter of a million prisoners and vast quantities of equipment were captured by the Allies. The final battles had seen the two Armies, the First and Eighth, each served by its own Air OP Squadron, fully employed on precisely those tasks which its pioneers had foreseen.

Invasion of Sicily and Italy

Next was the invasion of Sicily by 13 and 30 Corps in which one Air OP Flight each from 651 and 654 Squadrons took part. In this brief campaign of 38 days, there were two high points for Air OP: blowing up of an ammunition train by night harassing fire at Catarina railway station which was carefully registered by Air OP just before last light and Maj Neathercoat carrying out the first operational air shoot with the Navy on an enemy ship in Catania Harbour, which was unable to escape. This experience

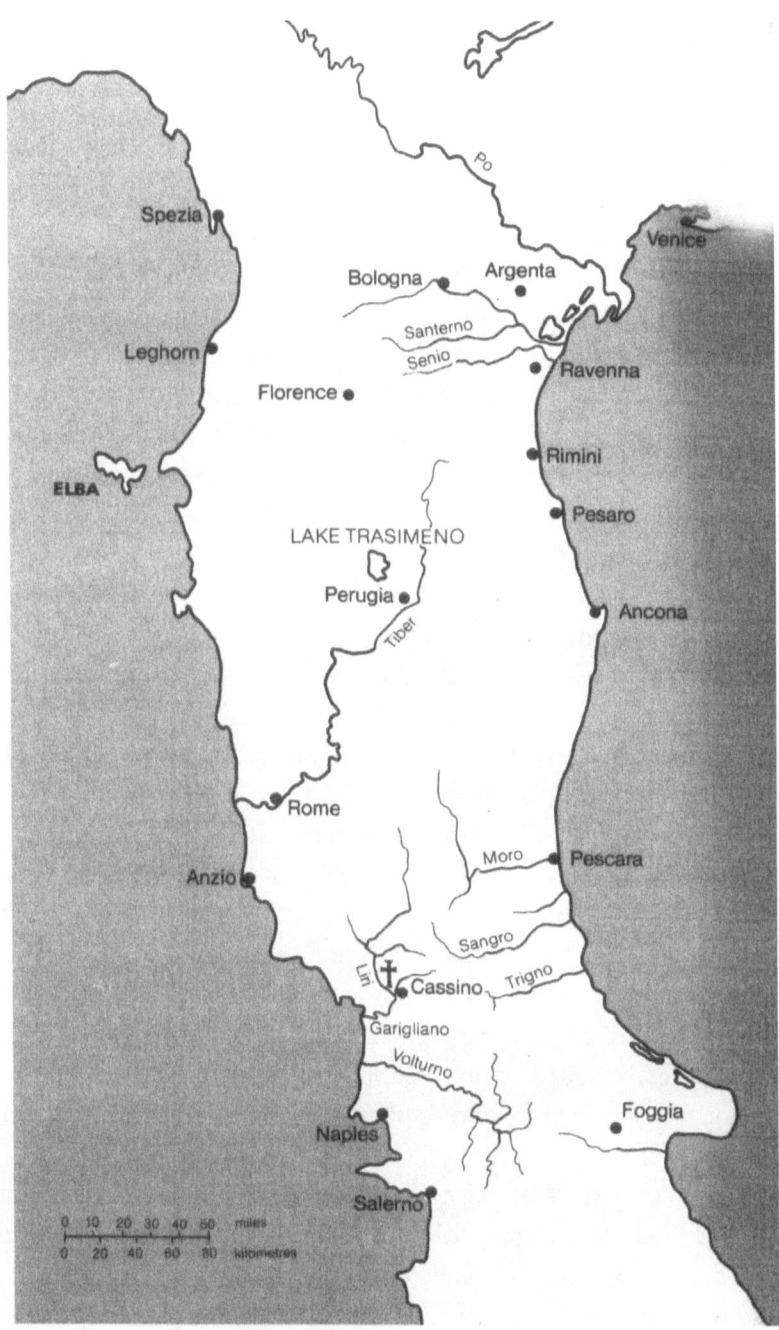

Map 4: Invasion of Sicily and Italy

was useful for future campaigns as later, whenever the Navy engaged land targets, Air OP was asked to do the observing for them.

The next big challenge and the process of the 'baptism by fire' for Air OP was the invasion of Italy where the terrain and hostile environment were very different from North Africa.

On D-Day, both 651 and 654 Squadrons started landing in the bridge-head at Salerno just south of Naples. Two pilots of 654 Air OP Squadron were allotted to observe the naval fire and successfully engaged the enemy armour with the guns of the US cruiser *Philadelphia*, forcing the enemy to call off its attack against the bridge-head. Later, during the advance to Naples, Air OP observed a large enemy formation of tanks which had not been located till then.

Air OP was in great demand in this campaign. About 90 percent of the registrations for the crossing of the Volturno were done by the pilots of 654 Air OP Squadron which between 9 September and 31 December 1943, flew 500 operational sorties.

651 Air OP Squadron was not far behind in its exploits. The *Times* correspondent's dispatch of 28 December said, "Some of the credit for the flexibility and accuracy with which our superior weight of guns is used must go to our Air OP Squadrons." The entry for 3 January 1944 in the 'A' Flight of 651 Air OP Squadron said, "The area has been under harassing fire from this time onwards. Houses on either side of Flight HQ were hit again today." Another entry a few days later said, "Captain Ward did 16 shoots today." Near Ortona, Capt Riley registered 19 shoots in one day, using gridded oblique photographs as a means of picking out his targets.

No doubt, Air OP pilots had become too bold, probably with the flush of success, followed by all round praise for their performance. So, on the New Year, a message was sent from the Army Commander, Eighth Army, both praising and warning them, *"We know from intelligence sources that the Germans have a great respect and dislike for the Air OP. They have, in fact, issued orders that no movement or firing will take place while the aircraft is in the air. It is possible that this success has made 10 Corps Squadron (651) forget the rules. Two days ago, I saw aircraft of 'A' Flight flying within easy rifle range of the enemy and I know that 'B' Flight do this continuously.......... I do not suggest that we should attempt to dampen the ardour of these excellent young men who pilot the aircraft but we are asking for trouble if we allow them to become too cheeky. Orders, therefore, should be issued that, except under exceptional circumstances and on direct command of a senior Royal Artillery officer, the aircraft should conform to the accepted rules of procedure and, thus, remain on the active list of their very important primary role."*

Meanwhile, 655 Air OP Squadron also arrived in Italy in January 1944, to support the operation of 6 US Corps, with British forces under its command. 6 US Corps had planned to use its superior sea power to open another front by landing at Anzio, south of Rome, to relieve pressure by the enemy holding up the advance of the British Eighth and the American Fifth Armies from the south. But unlike the landings at Salerno, this invasion encountered stiff opposition from the German forces occupying dug-in railway embankments and high ground behind the towns which was invisible to the ground observers. The terrain was also of ridge and furrow pastures where the enemy and his gun areas were engaged by Air OP pilots. The left flank where a German defensive locality was offering fierce resistance was engaged by Capt Peter Dutton with 6-inch salvos from the HMS *Orion*. During the next three months, 90 percent of observed naval shoots and 70 percent of all other artillery shoots were undertaken by Air OP pilots. As mentioned earlier, the success of the pilots could be judged from the fact that on the appearance of an Air OP in the area, the enemy ceased firing.

But this success soon earned the unwelcome attention of the Luftwaffe patrols which went for Air OP aircraft. In one such attack, Capt Dutton was shot down but on six other occasions, the pilots got away by taking prompt evasive action. Warning procedures and evasive action were further refined and paid dividends when in the next attack, Capt Gordon led his attackers to own light Anti-Aircraft (AA) guns which shot down one of the attackers.

In the first month of operation, 'A' Flight, with five pilots, flew 244 sorties and carried out 108 shoots.

In this campaign, there was a case of one Air OP pilot being hit by our own shells after which steps were taken to minimise such incidents.

Apart from the air shoots, the Air OP literally became the eyes and ears of the ground commanders performing other tasks, which contributed to the speed and success of the ground commanders' plans.

In one such case, Capt Fortnum observed a large column of armoured vehicles halted at a demolished bridge. He called up his regiment to find out who they were. They had no knowledge about this column and instead asked him to check it out. He flew very low along the column and saw that they were American armoured cars and in the absence of any communication with the column, landed there, praying and hoping that there would be no mines. An American Lieutenant came running across and informed him that they were the 91 Recce Battalion from Terracina and reported that the coast was clear of the enemy and asked the pilot to inform the main force with whom this column was not in communication. The pilot took off and through his regiment, conveyed the information. This was, in fact, the 'link up' which later became another task of Air OP in the fast moving battle, after the Normandy landings in particular.

On 26 July, Capt Cowley, Distinguished Flying Cross (DFC), from 655 Air OP Squadron had the honour of flying his Majesty the King from the operational strip at Radda to Siena – a flight of 20 minutes, with another eleven Austers carrying senior officers including Lt Gen Sir Oliver Leese, escorted by the Spitfires.

The Italian campaign ended on a happy note for Air OP. Maj Neathercoat, commanding 651 Air OP Squadron, ordered Capt Reynolds to proceed to Klegenfurt aerodrome in Austria and if it was serviceable (the Royal Engineers were expected to be there), to land. When Capt Reynolds arrived overhead, he saw a green verey light fired from the control tower. Presuming the Royal Engineers were already in control, he landed and on taxiing, saw ground staff signalling him in but as he went closer, he was surprised to see Luftwaffe personnel but still believed that they might be acting under the orders of the Royal Engineers. As he came out of the aircraft, he realised that he was the only Englishman around and the station had all Germans with weapons and a varied selection of aircraft. He was taken to the station commander and keeping his cool, he told, rather 'bluffed' him that he had come to ensure that there was no sabotage to their equipment and no harm came to his personnel and that his unit was to arrive there any moment. According to Reynolds, that was the most uncomfortable afternoon he had in his life, as by dusk, his story was wearing a bit thin when suddenly to his relief Maj Neathercoat appeared in his jeep. Reynolds told him what had happened. Maj Neathercoat quickly went off and returned half an hour after dark with some soldiers and posted them at strategic points, with machine guns covering the hangers and the runway. Maj Neathercoat and Capt Reynolds had shown remarkable presence of mind and 'captured' the airfield and the Prisoners of War (POW).

Normandy Landing

After heavy reverses in 1940, a small planning staff was put together to prepare plans to regain a foothold in Western Europe in due course of time. With victories in North Africa and good progress in the Italian campaign, this staff was increased to evolve a detailed plan which was refined by Gen Eisenhower and Field Marshal Montgomery. The final plan emerged for landings to be made on the Normandy coastline on 6 June 1944 (D-Day). The

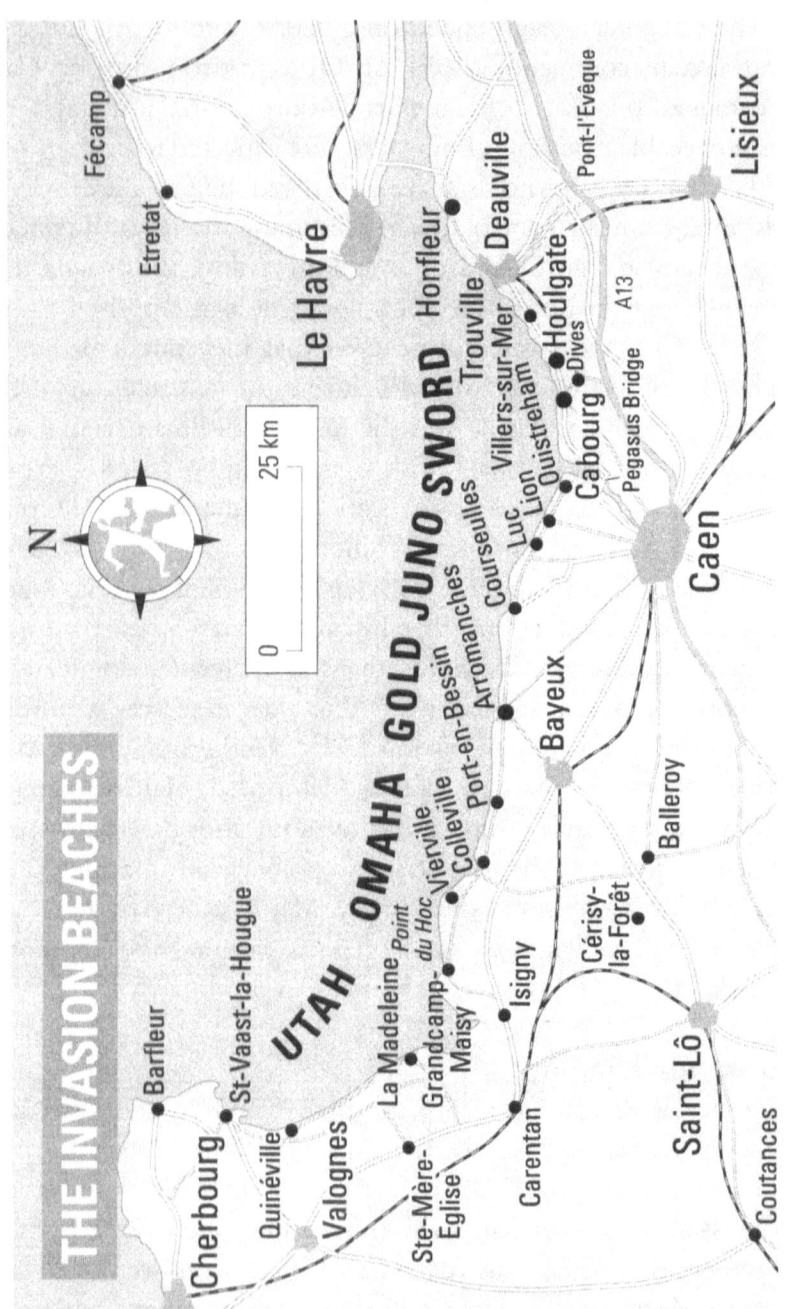

Map 5: Normandy Landing

assault was to be conducted in two phases: an air assault landing of 24,000 American, British, Canadian and Free French airborne troops shortly after midnight and an amphibious landing of Allied infantry and armoured divisions (1,60,000 troops along with their tanks, guns and ammunition) at 0630 hours. This involved an armada of over 5,000 ships consisting of 1,95,700 Allied naval and Merchant Navy personnel and hundreds of fighters and bombers of the RAF and American Air Force. This was the most complex operation ever undertaken in history. The details were meticulously worked out and rehearsals at appropriate level carried out. The secrecy of the operation as to the direction and place of landing was of paramount importance. Therefore, the destinations to the force commanders, at the appropriate levels were to be given at the last minute after which they were not to 'wander' about, described by some as "being put in prison."

Landings took place as scheduled along a 80 km stretch of the Normandy coast divided into five sectors – Utah, Omaha, Gold, Juno and Sword.

Air OP objectives and problems during such invasions were analysed in detail. These were availability of guns and ammunition which would be scarce initially till build up by the artillery in the bridge-head, reestablishment of radio communication between Air OP and guns, the air situation in the first few days (RAF senior officers were decidedly pessimistic about the survivability of Air OP), availability and preparation of landing grounds in the bridge-head which was expected to face the strongest resistance and onslaught by German firepower, both artillery and fighters/bombers and, lastly, navigation by Austers on the 96 km sea crossing. The solution arrived at was to send on D-Day, small reconnaissance parties by Air OP Squadrons with the ground troops to select and 'stake out' claims for landing strips and to fly the Austers in formations subsequently, led by Walrus amphibians of the Royal Navy which had the necessary navigational aids for flying over the sea.

A total of seven Air OP Squadrons formed part of this force. 652 and 662 Air OP Squadrons landed in the bridgehead on D+2 day followed by three squadrons on 13, 14, 18 and 27 June and two on 9 and 21 July 44. Later, in March 1945, 657 Air OP Squadron, also arrived from Italy in support of the Canadian Corps in Holland.

The first landings were on a strip 120 yards long in the bridgehead. The conditions in which they landed are described by a pilot thus, "Lester, David and Moffat were all there, looking very dirty but quite fit. It seems David had spent most of his time fishing people out of the sea!... Awful smell around here, hundreds of cattle over the place. Had some hot soup from a self-heating can, refuelled the kites and we all took off for a flight landing ground Jack Sullan had established three or four miles away towards Bayeux......Peter and I slept under our 3-tonner, Terry under his aircraft, Jack, Mac and the boys in holes and ditches. Soon after dark, a big firework display all along the beaches. Boche (German) kites beating up the shore parties and landing craft. All night continuous streams of tracer going up...fell asleep watching it. We are four or five miles inland......."

The air observation and control of fire of naval guns for the Armada on D-Day and for a couple of days after was provided very effectively by high performance aircraft of Naval Aviation but as the bridge-head enlarged, for many reasons, this could not be continued indefinitely. The early deployment of Air OP units in the bridgehead ensured the continuance of observation when the Air OP pilots also directed the firepower from these powerful floating batteries. On 7 July, the day before the final assault on Caen, the first town in the north of the bridge-head, Capt Cobley carried out a shoot with the big guns of the HMS *Rodney* onto targets just north of Caen. An eye-witness account of this shoot, by a strange coincidence, by Lt Col HC Bazeley, the father figure of Air OP, who by some stroke of fortune was commanding a Light AA Regiment deployed at a place in the bridgehead which saw the greatest air activity in the first 40 days of the campaign, read, "On 7 July......I watched Cobley carry out a shoot with big guns of HMS *Rodney*....... He operated from immediately above the deployment area of Fox Troop who were in wireless touch with him from a set situated at their warning gun. As I listened to his orders over the wireless and the response from the Navy and the whine of Rodney's big shells as they passed overhead, I looked back on the years since the summer of 1938 when the Air OP had been only a dream. Safe in the security of its own AA defences, it was controlling the biggest guns of His Majesty's Navy – that shoot seemed to me the epitome of all for which we had striven." It was a dream fulfilled for Bazeley who was emotionally moved to see this air shoot.

Bazeley's AA regiment had also perfected such an efficient system of warnings, look out and protective fire that despite repeated attacks on the Air OP in that sector, only one aircraft was lost and that too without the loss of its crew.

This was what Von Rundsted the German Commander said about his unsuccessful efforts to contain the Normandy bridgehead: "The flexible and well-directed support of the land troops by the ship's artillery of strong English naval units ranged from battleships to gun-boats ...... the enemy had deployed very strong naval forces off the shores of the bridge-head. These can be used as quickly mobile, always available, artillery. *During the day, their fire is skilfully directed by air observers and by advanced ground parties.* Because of the great rapid-fire capacity of naval guns, they play an important part in the battle within their range. The movement of tanks by day in open country within naval gun range is scarcely possible." This was a tribute by the German Commander to both the naval aviators and Air OP pilots.

So intense was the air activity from both sides in the bridge-head area that leisurely flying by the Air OP above 1,000 ft was rarely possible. Therefore, another classic method, popping up, was evolved whereby the Air OP pilot flying at tree-top level, pulled up in steep climb to synchronise his observations with the fall of shot, transmitted the next order and then swiftly swooped down to tree-top level, to wait for the next round.

Another account by Maj Cobley, commanding 652 Air OP Squadron in support of 51 Highland Division, said, "In support of the operation, 57 sorties were flown, including 49 shoots...... more than 27 enemy aircraft tried to interfere. Captain Bawden and his rear observer, Gunner Passmore were shot down by five Me109s. They made a safe landing, though Passmore was wounded. These five Me109s patrolled the area for an hour and drove the Air OP away, but they never pressed their attack once the pilot had seen them and begun to take evasive action. Three Me109s were shot down by our flak (A A guns), one by Spitfires."

After the D-Day, it took a number of days for Allied artillery to build up to full strength when larger concentrations of guns were turned onto important targets. On 17 July, Maj Lyell Commanding Officer (CO) of 658 Air OP Squadron conducted a shoot on 40 enemy tanks lurking under cover. *In this, the artillery of 12, 30, 2 Canadian Corps with their supporting Army Group R A of more than 500 guns of field, medium, and heavy artillery participated. This was perhaps the largest concentration ever controlled by an Air OP pilot till then.*

The sterling role played by the Air OP units, whose survivability was in doubt in the bridge-head battle in the planning stage, was best summed up by the German 10 S S Panzer Division, a very formidable formation, *"But the greatest nuisance of all are the slow-flying artillery spotters which work with utter calmness over our positions, just out of reach, and direct artillery fire on our forward positions."*

Breakout and Pursuit

Having failed to contain the bridge-head through June and July, the final German counter-offensive against the Americans, launched on 6-10 August, was also unsuccessful. This resulted finally in a great and dramatic break-out by powerful armoured columns on the right of the bridge-head.

Maj Hill, commanding 662 Air OP Squadron, described the lightning break-out and advance of 30 Corps (commanded by Lt Gen Sir Barian Horrocks) and noted that for days on end, the flights were completely out of touch with Squadron HQ except for passing information about their location over the Corps Command net. A pause in the advance also gave them no respite as the Air OP pilots were kept busy flying all day long between Divisions and Corps HQ, being the only means of contact and communication between HQ and forward troops.

The entry in the diary of 'B' Flight 653 Air OP Squadron of 18 August states: "An epic day. The gap practically closed and enormous slaughter of enemy tanks, MT and infantry carried out by the Division... 'B' Fight occupied new landing ground at 0830 hours and carried out almost continuous sorties all day. Coles.... flew six hours forty minutes and did twelve shoots from Battery to Divisional Targets. His grand total was six Mike (Regimental) targets, five Uncle (Divisional Artillery) targets and one Battery target. All on tanks or MT Columns ... By evening, the whole area was littered with fires from burning vehicles, tanks and woods."

Air OP faced strange situations and adventures like the one in the Italian campaign as mentioned earlier when an Air OP pilot had landed at Klagenfurt airfield in Austria still under German control.

'B' Flight, in support of the Guards Armoured Division, after arrival in Brussels on a Sunday night, was not able to find an ALG the next morning and decided to use Brussels civil airport, presuming that it must be all clear of the enemy. The Flight Commander flew there to have a look, made low runs up and down and finding no signs of life, flew back, dropped a message on Corps HQ saying, "The airport was OK" and he was going to use it. On arrival back at the ALG, he ordered four Austers to go and land on Brussels airport, and himself got into a jeep to reach there by road simultenously. As he got there and the four Austers, after lowering their flaps, committed themselves for landing, the fun started. A battery of 88 mm German guns exactly 1,100 yards from the edge of the airfield, lying 'doggo', opened fire. The aircraft taxied crazily and got behind the two hangers. Only one aircraft was hit but without injury to the pilot – a miracle. The situation was that behind the two seedy looking hangars were four aircraft, four pilots and the Flight Commander in a jeep and a German battery of 88 mm firing over open sights at the hangars – "very unpleasant indeed," with the undamaged aerodrome worth in gold for the grab. During all this mess-up, the Squadron HQ transport also arrived, to be greeted by a hail of fire by the German battery. A message (couched in rather excited tones) to the Corps HQ managed to get through and the answer was, "Sixteen armoured cars and a company of Guards sent." By nightfall, the guns had been captured. Casualties comprised one aircraft, 11 trucks damaged but serviceable and no personnel hit and the gain was an aerodome intact. Air OP Squadron operated from there for about a week and when it moved forward, the Dakotas were coming at the rate of 40 an hour and a wing of Spitfires had also started operating from the airfield. Nothing could be more gratifying to the personnel of this Air OP Squadron who carried sweet memories of "capturing the airfield intact" for days to come.

Later, 'A' Flight with 50 Infantry Division which were left miles behind in the mopping up operations, had less problems of movement but got plenty of shooting. 'B' Flight moved up with the Guards Armoured Division and landed itself in another unpleasant situation when one morning at 5 am, it found two Panther (German) tanks on the ALG. The tank crew did not notice the Austers which were parked for the night under some trees, with the personnel of the flight having gone into a broken down factory alongside to sleep. The tanks saw the lone guard, opened up, and after firing for 10 minutes, turned around and drove off. No damage was done.

'C' Flight was not to be left behind. Its ground parties fought a pitched ground battle when, during their advance, they encountered an enemy pocket just west of Antwerp. The Flight Commander sent a message asking for a tank to come up. Together, they captured 36 prisoners, including two officers for the loss of one soldier killed, which was a pleasant memory for the flight personnel for years to come.

During this rapid advance came the demand for communication flying for commanders and staff which at times was too exciting to be pleasant. In September 1944, Capt Stunt with Brig Jack Parham, Brigadier Royal Artillery (BRA) (one of the founder fathers of Air OP) and Capt Salter with Brig Chilton, Chief of Staff, both from HQ Second Army, flying in two Austers, were heavily engaged by 20 mm and small arms fire. Stunt and the BRA managed to get through intact but Salter's aircraft was severely damaged and he had to crash-land. Fortunately, neither he nor the Chief of Staff was hurt. The Chief of Staff sat in a ditch and burnt his secret documents with a cigarette lighter, expecting the worst, but he and Salter, guided by a Dutch farmer, made their way back safely to their forward localities.

The Canadian troops once made a demand for a new task for the Air OP—aerial control of traffic. An Auster from 660 Air OP Squadron was fitted with a loudspeaker. The Air OP pilot detailed for the task successfully unravelled some complicated traffic jams.

Another encounter with enemy fighters and 'classic' Air OP evasive action witnessed and described by Maj Hill, commanding 662 Air OP Squadron, goes thus: "Captain Cracknell on getting the warning of 'bandits,' came down very low for ten minutes but seeing three Spitfires (own fighters) flying above him, climbed up to continue his registration shoot. Unfortunately, above the Spitfires, flying in and out of the clouds, were two FW 190 German fighters which dived and went straight for Cracknell who seeing them, dived for the ground. One of the Spitfires also dived and shot down one FW 190 but crashed himself. For the next 10 minutes was the battle between the Auster and the FW 190 .... Round the trees, down the side of the river, round the chimneys of the power station. The FW fired four bursts but missed each time and then cleared off on seeing the AA getting too hot for him. Cracknell landed safely but the Auster had to be completely re-rigged, one wing having gone back about an inch during the attack...... I spoke to the rear observer some two minutes after landing. He was a young Irish operator-driver. When I asked him if he was OK, he said, 'Oh yes, Sir, quite – Captain Cracknell is a very good pilot'."

Siege Operations

After break-out from the bridge-head, the Allied Army detached 1 British Corps and the 2 Canadian Corps, both under the First Canadian Army to deal with the German coastal garrisons left at Le Havre, Boulogne, Calais and Dunkirk. The object of these garrisons was to deny the Allies use of these harbours for landing of additional forces and weaken the thrust of the Allies towards the German mainland. These ports were also the launching sites for V-1 and V-2 rockets which were still being fired on the British mainland. Le Havre was the first to be assaulted, where the harbour rested on a peninsula, well fortified and one of the strongest fortresses of all. The attack started at 1745 hours on 10 September after the defences had been softened over the preceding few days by fire from the Royal Navy, artillery of the attacking 1 Corps and bombardment by the Bomber Command which in the last 90 minutes before H-hour, dropped over 5,000 tons of bombs. The operation was over in 48 hours, resulting in the surrender of 12,000 soldiers.

Pilots of 652 Air OP Squadron and two flights from 661 Air OP Squadron, with Maj Cobley, CO 652 Squadron as coordinator of the Air OP support, flew over 500 sorties during this operation. The task of Air OP was to locate and engage 64 enemy gun positions, including their 'flak' batteries, particularly during the raids by the Bomber Command.

After this battle, a German Artillery Commander of Le Havre, when asked why the many Austers flying around were not engaged by his 'flak', replied that *the fire which the Austers turned on in reprisal was so severe that his Battery Commanders considered the game not worth the candle.*

Next was the assault on the garrison of Boulogne which was defended by over 10,000 German troops, supported by more than 90 guns ranging from 75 mm to 350 mm under the command of Lt Gen Heim who was determined to hold the port as long as possible.

The artillery plan for the operation 'well-hit' was a complex and sophisticated document that coordinated the contribution of 368 guns, including heavies and two anti-aircraft regiments firing airburst in a ground support role. The guns were supposed to neutralise the enemy forward position and strong points as well as the German artillery. Over 400 targets to be engaged by predicted fire were to be corrected by Air OP pilots after the fire plan commenced.

The assault on Boulogne led by 3 Canadian Division was supported by Air OP units of 660 and 661 Air OP Squadrons. It commenced on 17 September and ended on 22 September with the surrender of 9,500 Germans – officers and men.

A very unusual episode in this battle was when Capt P Hawkins of 661 Air OP Squadron was wounded in his right hand, arm and shoulder from ground small arms fire. His rear observer Lance-Bombardier Gibbs, an airman, saw Hawkins gradually losing consciousness, and with great coolness and presence of mind, he undid his safety harness, came up from the aft-facing rear seat and leaning over Hawkins, managed to control the aircraft and, at the same time, rendered first aid to him. Hawkins eventually regained consciousness and although in pain, was able to handle the controls but only with his left hand. He also found it difficult to operate the 'stick' and the throttle at the same time. With joint operation between the badly-wounded pilot and the observer, the aircraft was landed successfully. Gibbs was awarded the Distinguished Flying Medal (DFM).

Another high and defining moment of the assault on Boulogne for the Air OP undoubtedly was the observation of the fire of the super-heavy long-range guns emplaced near Dover on the English mainland, firing across the Channel to support the Canadian attack at a range of 42,000 yards. The time of flight of the shell was about 80 seconds. It required elaborate briefing and assured long range wireless communications. First shoots, mainly experimental were carried out on 16 September. The next day, three targets were engaged by Air OP during the actual assault in which a direct hit was scored on the enemy gun emplacement, throwing the gun off its mounting. Then, on 18 September, further air shoots were undertaken, including on a 'flak' battery which had been bothering the Air OP pilots.

On 19 September, Air OP pilots, due to still heavy 'flak' fire resorted to tactical flying, flying extremely low and then pulling up rapidly to synchronise the observation with the fall of shot. Sixty-two rounds were fired that day with good results from the super heavy guns. The next day, the guns were no longer shooting together as was inevitable with high velocity long-range guns whose barrel life is very limited. Altogether, 200 rounds were fired in these cross-channel air shoots.

The assault on Calais began on 25 September and ended on 30 September when the garrison capitulated, resulting in a haul of about 10,000 prisoners. Dunkirk, with 15,000 German soldiers, was the last to fall in the due course of time.

The Rhine Crossing

The crossing of the Rhine planned for 24 March 1945 was the last battle of the war as its successful outcome would place the Allied Army in the heart of the German mainland. This was to

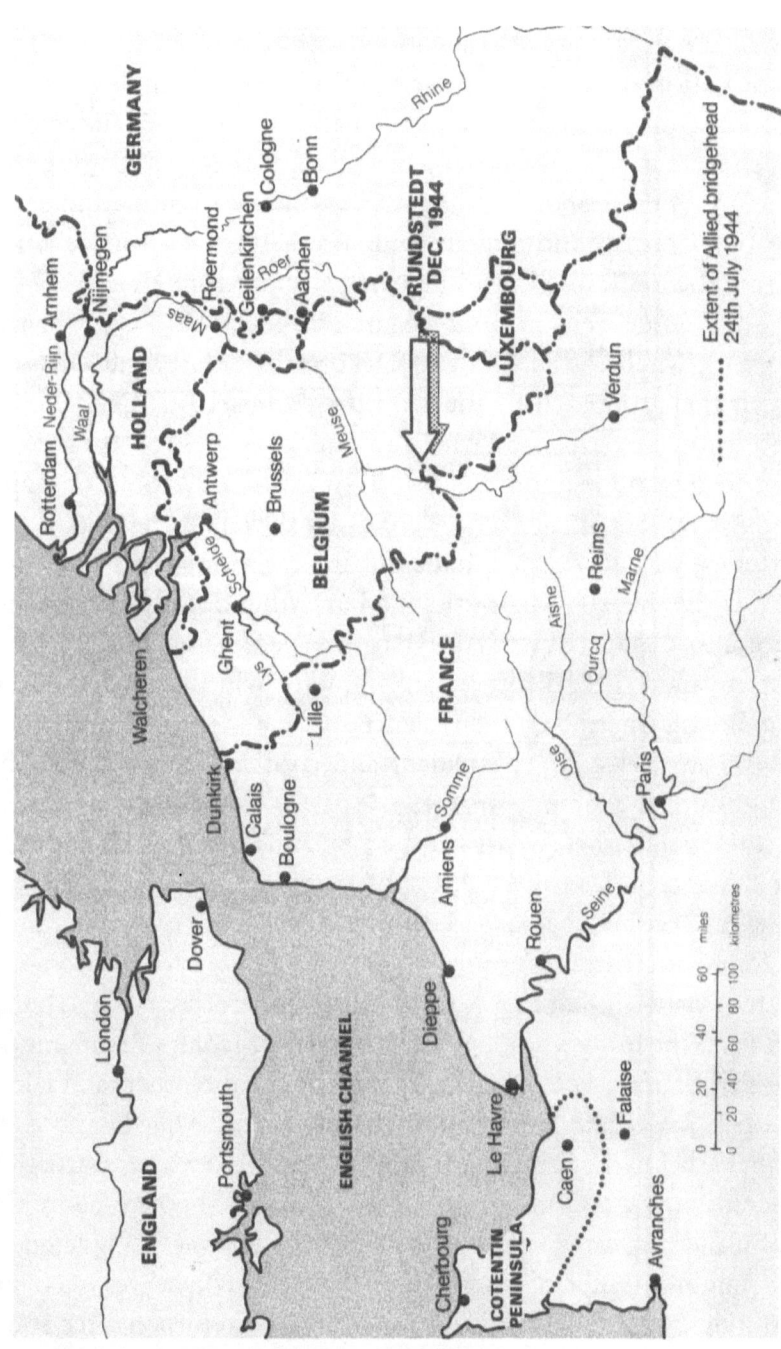

Map 6: The Rhine Crossing

be an Anglo-American operation with the Canadian Army's participation on the extreme left and large scale employment of airborne forces comprising a vast armada of aeroplanes and gliders which were to take place over a period of three hours. During this time, many thousands of Allied Army guns were required to cease fire but German anti-aircraft guns firing on the airborne force were to be neutralised. This task was given to Air OP.

The assault went in on 24 March 1945 at 10 o'clock in the morning when the gliders and Dakotas of 6 Airborne Division began their approach to the Dropping Zones. An Air OP pilot who witnessed this and the subsequent waves of paradrops, described it as "incredible." Air OP pilots of 652, 653, 658 and 661 Air OP Squadrons who were given the job to neutralise flak batteries, carried out a number of impromptu shoots on some very aggressive 20 mm positions. Many other opportunity targets were also engaged by Air OP. In the space of an hour, a pilot first engaged three Panther tanks, one of which had already hit two British Shermans, then silenced three enemy SP guns by a salvo from 7 Medium Regiment and another enemy battery by the guns of the same regiment.

Targets like these were engaged by Air OP pilots throughout the day and on D+1 & D+2. Capt Cover of 'B' Flight 658 Air OP Squadron on D+2 had a field day when, in a space of two minutes, he found two batteries consisting of three guns each and two tanks in an orchard causing a lot of trouble to Allied troops trying to break out from the bridge-head. Cover ranged 64 Medium Regiment on the first battery, 121 Medium Regiment, on the other, and laid a troop of 7 Medium Regiment on the two tanks. While he was busy in these shoots, he spotted an ambulance painted white, evacuating casualties from a farm house. He stopped the shoot to allow the ambulance to carry out its evacuation which was completed in five minutes. After the ambulance disappeared, he resumed his shooting but, unfortunately, this ambulance returned after some time for another evacuation and was inadvertently hit when Cover was engaging the retreating enemy tanks. He was awarded the well-deserved DFC for his commendable work in this battle.

With thousands of guns, from 40 mm to 9-inch, firing according to a pre-arranged programme as well as on the direction of ground and Air OP, the risk of Air OP aircraft being hit by own guns could not be totally ruled out. While three Air OP pilots and their rear observers were killed that day, another pilot had a miraculous escape when a 5.5 inch shell passed through his rudder without exploding or altering the trim of his aircraft.

For the next two days, the Air OP pilots found fewer targets but were kept busy on reconnaissance tasks after the Rhine was crossed. The citation for the DFC to Capt Hargreaves of 653 Squadron is an example of one such sortie *"In April 1945, this officer was detailed to make a reconnaissance of the bridges over the Dortmune-Ems Canal which were held by the enemy. This entailed flying, in some cases, directly over the bridges and in the face of intense enemy opposition. His aircraft was hit several times and damaged; nevertheless, Capt Hargreaves completed his mission successfully and the information he obtained was of the greatest value to the Army. Later, in the same month, this officer was again detailed to reconnoitre twelve bridges over the river between Alfhausen and Bramache. The weather was very misty and the bridges were 5,000 yards in enemy territory, but after making sorties lasting two and a half hours, Captain Hargreaves reported their condition accurately and, as a result of the information he obtained, the Divisional plan was completed many hours before expected and much patrolling was saved."*

Captain Wawn of 652 Air OP Squadron has given an account of his last operational flight: "After the liberation of Apeldoorn (Holland), I was detailed to go with a section to work with 185 Brigade of 1 Canadian Division. This Brigade was acting as an independent Brigade. On the previous day and this day, they had been pushing on trying to contact the Boche (Germans), who had retreated very rapidly. I was called on to do a road reconnaissance of the proposed Brigade route, see if it was suitable and if there were any signs of opposition. Since at one point the prepared route was impossible for vehicles to use, owing to its being very overgrown, I circled round over the village of Hoenderlo, at about 3,000 feet, to plan out an alternative route. The only alternative route lay through this village and on inspecting it, I noticed a road block and lots of people whom I took to be waving. As it was unlikely that civilians would be around in numbers if the Germans were there, I dived down to about 1,000 feet. They all appeared to be civilians, so I decided to drop my card. I wrote out some messages, telling the civilians to open the roadblock and, going right down, I dropped the message on to the crowd. The people seemed almost mad with excitement. Returning, I dropped a message on the road column, telling them all about the likely change of route and the roadblock. The wireless link had failed owing to the distances, and the Brigade continually moving. I reported to the Brigade on landing."

"Later in the day the Brigade made its HQ in Hoenderlo, and after landing, I reported for orders. To my astonishment, the Brigadier stopped me and said, 'I must congratulate you on the capture of these forty gentlemen', pointing to the German prisoners. Apparently the villagers on receiving my message had not only pulled down the roadblock, but had rounded up these Germans who had dropped out of the retreating columns and were hiding. They handed them over to the first Canadian officer they saw."

With most of the German hinterland occupied by the foreign Armies including the Russian Army, fighting came to an end at midnight on 8 May 1945. But this was followed by indescribable chaos and the occupation Armies got involved in restoring some order in the prevailing chaos. In addition, it was a stupendous task to feed hordes of human beings, numbering many millions, compressed mostly into the northwest corner of Germany where they had fled to avoid the 'dreaded' Russians. Thus, Air OP became even more busy in communication work to fly Commanders and staff who could reach destinations in minutes instead of hours by the roads which were in bad shape.

South East Asia

Now let us move onto the Burma front where Air OP faced entirely different battlefield environments and challenges, altogether different from the North African or European campaigns.

Only one railway line ran roughly parallel to the front with only three rail-heads – Chittagong for the Arakan sector, Dimapur for the Manipur sector and Ledo for the northernmost sectors. Roads from the railheads to the outpost had to be built through swamps, very dense jungle and along the precipices of mountain ranges.

The Allied troops also faced an enemy who was fanatical to a degree not seen elsewhere, since surrender was not part of his code of war. The Air OP operations further faced a peculiar set of challenges.

The first and foremost was the weather. The monsoon from April/May to September/October brought torrential rain, swirling clouds, violent atmospheric turbulence which made flying dangerous and highly unpleasant. Under these circumstances, it was also estimated that the aircraft fabric would only last four months and wooden propellers and the perplex canopy would also need replacement after about the same time. The nearest repair and service unit was at Calcutta.

Another major problem was the vast distances from the railheads to the troops on the front line. Flights could be 150 to 400 km away from each other and Squadron HQ, located at a central place, had to fly the spares and technical stores to the flight which otherwise would take a few days by road.

Finally, a major problem was how to cope with the load of work as only one Squadron was allotted for the whole of 14 Army.

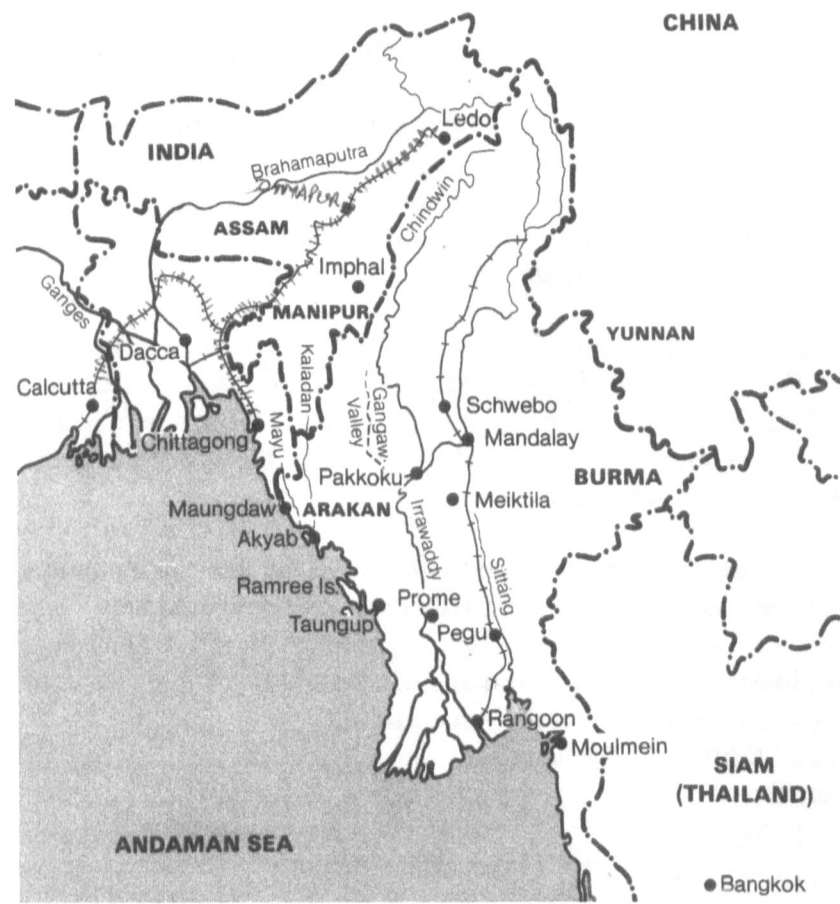

Map 7: South East Asia

Thus, a Corps was allotted only one Flight instead of a Squadron consisting of three Flights, with the result that a Division was supported by only one pilot instead of a complete Flight.

With this backdrop, 656 Air OP Squadron which sailed from the UK on 14 August 1943, arrived in Bombay in September 1943. The first batch of Austers was assembled at Bombay during October and November. Another consignment of aircraft reached Calcutta in December 1943.

The squadron then moved from Bombay to Deolali. On 12 January 1944, the squadron less 'B' Flight which was allotted to 15 (Indian) Corps on the Arakan front, moved from Deolali and arrived in Calcutta on 22 January 1944. Arriving in the Corps sector on 28 January 1944, it went straight into action. In the very first sortie ordered, due to breakdown in communications, Capt McMath noticed that the attacking tanks were held up but the timed artillery programme was continuing as per the fire plan. Having no message bag, he dropped a message by putting it into his sock which landed at the door of the Commander Royal Artillery's (CRA's) *basha*. McMath later 'explained' and 'apologised' to the CRA that he was sorry as he had been wearing the sock for a week!

On 3 February, when the Japanese launched a big offensive, Capt Boys, during a reconnaissance flight, was shot down. Badly wounded after crash-landing, he was saved by the Burmese who carried him through the Japanese lines to safety. On the same day, Maj Coyle, the Commanding Officer, and Capt Maslen-Jones flew several sorties to the most isolated forwarded posts, at times landing on small strips, wading through hostile fire to deliver urgent operational orders and medical supplies.

On 23 February, 7 Indian Division and also the gunners and airmen of 'C' Flight had been cut off in the famous 'Admin Box'. In mid-April, the Squadron and 'C' Flight were ordered away from the Arakan to Kohima in the Manipur sector, leaving 'A' Flight in support of 15 Corps. On arrival at Dimapur, they were placed under the command of 33 Corps for the relief of Kohima which came on 22 June.

'B' Flight which had been left behind at Deolali, was ordered to move and join 4 Corps at Imphal. It left Deolali on 16 February with four Austers which were flown from Deolali, arriving in Calcutta a fortnight later. As nobody would take the responsibility, the flight carried out this move from Deolali as a normal tactical move on its own initiative. Since very few known landing grounds were available on the 1,500 miles route to Calcutta, the four Austers, in two pairs, would follow the road, choose the strip and land, if found suitable, and inform the ground party to reach there or instruct the ground party to go ahead and prepare the strip where the Austers would land subsequently. From Calcutta, the Austers flew directly to Imphal on 9 March and the ground parties reached a week later, covering 540 miles by rail and the last 120 miles by road.

One of the first orders given to the flight was to send two sections to join 20 Division in the Kabaw Valley, some 25 miles east of Imphal. 20 Division at this time was withdrawing according to a pre-arranged plan. At the same time, 17 Division was also falling back on Imphal, with 23 Division already in the valley. Imphal was cut off when 5 Division was flown in to reinforce these three divisions. Fly-in involved about 300 Dakotas, Skymasters and Wellington aircraft coming in daily. 'B' Flight was now busy supporting these four divisions, one in each corner of the valley. With the outbreak of the monsoon and the flooding of ALGs in the division, the flight had to fall back on the two main all-weather strips at Palel and Imphal, which were being used to fly in the reinforcements. With half a dozen Dakotas circling overhead and about the same number waiting to take off, the incoming Austers, often needing quick refuelling to return to resume their registration of targets, would not get priority during such heavy air traffic. On one such occasion, Capt Fowler, who returned to refuel after suspending a registration shoot in which he was to adjust each battery of the divisional artillery on the target, was not able to land, with the Dakotas getting priority over him. Already low on fuel, he fired a red verey light on the air traffic control after which he was allowed to land. After refuelling, when he was not able to take off, as the empty Dakotas were given priority, he stormed into the air traffic control and shouted at the traffic control officer saying that 72 guns were waiting for him to engage the Japanese and "couldn't empty Dakotas wait"! From then on, all the Austers had priority to take off and land over other aircraft except the Spitfires on fighter patrol missions.

In addition to the normal air shoot and information sorties during this 'siege' in which the Japanese would close in every day, Air OP undertook various additional tasks like air drops of supplies to ground OPs and outposts on hill tops (normally a two-day journey on the ground), scattering of propaganda leaflets, map making/drawing of sketches which were later mass produced for use by the infantry.

Mention must be made of some narrow escapes and lifetime experiences of some pilots (at least to educate those in our Army who think that Air OP is a sitting duck). Such instances also show the initiative, daring and fearlessness of their character and the stuff that Air OP pilots are made of. Capt Southern, for instance, one morning set off on an outlying airstrip where a ground party was to meet him. During the night, the Japanese, dressed as civilian refugees, infiltrated there, surrounded the area and cut the lines of communication of the Allies. He circled around the strip at about 800 ft. He saw no signs of the ground party or the Japanese but saw a lot of mules wandering loose in the nearby paddy fields. Presuming them to be 'friendly' mules, he decided to land in full view of the Allied troops who were certain of his capture or death and destruction after landing. As soon as Southern switched off and came out of the Auster, he saw grenade-dischargers and mortar bombs falling on the strip, with tank and small arms fire coming from another direction. Keeping his nerve, he poked his head into the cockpit, put the fuel and other switches 'on', and then swung the propeller. The engine fired but in the hurry, as he had forgotten to ensure 'throttle fully back', the Auster started moving forward. Ducking under the wing, he dived into the cockpit and took off, with his feet still dangling out of the door. This spectacle was narrated by some Allied troops who watched his providential escape in agonised silence.

By May/June, the monsoon was in full fury and operational flying became almost impossible. The aircraft were also showing serious signs of deterioration and the whole squadron was withdrawn to Ranchi for a refit where it stayed till September 1944.

In September 1944, 'C' Flight moved from Ranchi for the Arakan front in support of 15 (Indian) Corps consisting of 15 Indian Division at Cox's Bazaar, 25 and 26 Indian Division at Maungdaw, the Commando Brigade on the coast and 81 West African Division in the Kaladan Valley. 26 Indian Division was later replaced by 82 West African Division. The flight also observed for the Navy having two Australian destroyers and a 6-inch gun cruiser and later two Indian sloops. This Corps also had an independent force of around with 7.2 Howitzers and a troop of 25-pounders. Thus, a single Air OP Flight had to provide support for the 15 Corps of four divisions, the Royal Navy and two independent brigades. The flight at one time grew to seven pilots and seven Auster aircraft.

'A' Flight was soon actively engaged, supporting the 11 East African Division, the leading division of 33 Corps which was advancing in very thick jungle along the only road axis running along the north bank of the river Irrawaddy. On both sides of the river, great ridges projected almost vertically, rising 1,000 ft, affording perfect observation and machine gun positions, while the jungle was so dense that it was not always possible to see the ground from the air and infinitely worse and impossible for the ground troops to locate anything in front till a hail of fire greeted them. At times, it was hard for the Air OP pilot to even spot own forward most troops. Thus, the country was ideal for the defence and horrible for the attacker.

To make the best of the worst situation, an aircraft was kept in the air most of the day in direct Radio Telephony (R/T) communication, with the gunner officer with the forward most company commander. On encountering the enemy, the pilot was informed and indicated the forward most position of the company by putting up a big orange umbrella which showed up proficiently among the green trees and undergrowth. Then the pilot would come down low to fly up and down, searching for signs of the enemy's movements or its bunker positions.

Even if the enemy was not spotted, a few shells were directed on their likely positions. At times, this worked as the enemy, now retreating, could be seen falling back. Realising the impossibility of locating the enemy's strong points, the General Officer Commanding (GOC) ordered that such likely targets should be registered early so that fire could be brought down on any point without delay.

Air OP pilots were often sent to locate and contact the forward brigades/isolated troops in the event of a breakdown of communication between them and the Division HQ. During these sorties, impromptu air shoots with the divisional artillery were also undertaken whenever the Air OP pilots located the enemy movements.

By now, the Japanese had given orders to their troops to stop all movements and hold the fire of their guns on sighting an Air OP aircraft. This advantage was exploited to protect the Dakotas flying low on supply dropping missions which were regularly engaged by the Japanese guns. An Air OP aircraft was flown just above the Dakotas, with the result that the Japanese stopped engaging them. Thus, the Dakotas flew off home safely after dropping their loads.

As the road journey was out of the question, the commanders were often taken for the reconnaissance sorties by the Air OP pilots. Gen Rees, GOC 19 Division, frequently went up in the Air OP aircraft for reconnaissance over the battlefield and would insist on crawling into the rear seat to enable the pilot to take the wireless set which could only be tied on the front seat, so that he could keep in touch with the artillery regiments to engage any opportunity targets.

In another sortie, Capt McMath was tasked to find out the exact whereabouts of an infantry company which had secretly sneaked across the Irrawaddy to prepare the way for the large scale crossings later. This sortie had to be carried out in such a way that the Japanese suspicions would not be aroused. Capt McMath flew very low over the river, all the time turning and twisting in a carefree manner till he spotted some troops on the top of a cliff waving at him. He returned to the division with the good news.

Capt McMath who was the only pilot supporting 19 Division has described the type of work he was engaged in as under:

> One of my main jobs during this period was searching for enemy guns which had been located approximately by means of flash and sound bearings. At other times, I had to patrol over the whole front, hoping that I might spot a gun flash and so crash some of our medium artillery on to it. Although the Jap never used his guns with anything like the effect of the Royal or Indian Artillery, he was obviously bringing up a greater concentration than anything we had met before. As January advanced, our forward troops and even the Brigade and Divisional Headquarters suffered casualties and frequent harassing. On the face of it, I should have been able to spot his guns, at least when they fired, and so destroy them one by one. But I'm afraid that in practice, the 'Hostile Battery' quest – as it was officially called – proved to be my biggest failure. When not actually firing, the Jap guns were, I believe, impossible to see, so effective was their camouflage and use of natural cover. So strict was the discipline of

the men forming the detachments that they must have remained stock-still for long periods while I flew to and fro over their heads peering down through binoculars at the suspected position which I could so rarely verify. Occasionally, some obvious wheel-tracks raised my hopes, until I followed them to find them disappear into a blank wall of jungle, or even peter out in open paddy where not even a sniper, far less a gun with its detachment, could have been concealed. Still more of an occasion was the discovery of something that looked vaguely artificial at the end of the tracks, whereupon I would always have a shell or two to clear away any undergrowth and possible camouflage, and so to reveal whether I had been made a fool of again or whether I really had discovered something at last. Very rarely, a black gun-barrel or the spokes of a wheel would be revealed beneath a great tree as a cleverly placed bush was blown away by a bursting shell. Then, with a whoop of delight, I brought all our own artillery crashing down until nothing remained but a tangle of broken branches and torn up earth, with a few metal fragments shining duly in the sun. If this sort of success didn't happen often, at least when it did, it was the most exhilarating experience one could possibly desire. Once the Jap had learnt the un-wisdom of showing himself in daylight he reserved his day's shelling almost exclusively for the periods while I was back at Onbauk, refuelling and feeding. At another time, Dakotas flying low on their supply-dropping circuits were fired on by Japanese field guns pointed upwards as AA weapons, and so for the next few days, I always flew just above the Dakotas to stop any firing – but hoping, I remember, in my heart that they might commit just one indiscretion. But they didn't, and the Dakotas dropped in safety and flew off home.

A small part of my duties with 19 Division was flying the GOC when a road journey was out of the question, or taking him for a reconnaissance flight over the whole battlefield. Late one afternoon, after we had flown fairly deep into enemy territory to study the terrain over which the division might soon be advancing, I was amazed to see a field gun right in the open, firing several rounds as I flew straight overhead. It was right in the middle of the ground over which I had been daily searching in vain for Jap artillery. Nothing has ever given me such a feeling of impotence as sitting up there with no wireless in the aircraft, and so the little Jap gunners went on pumping shells across the river at the British and Indian soldiers. As a result of this one 'get away', General Rees always insisted in future on crawling into the rear of the cabin so that the wireless could occupy its usual place by my side and enable me to maintain constant touch with an artillery regiment throughout our flights.

It was evident that only a gunner officer with sound professional knowledge of the ground battle could perform as McMath did, and he was decorated with a well-deserved DFC.

During this campaign, Air OP pilots flew some of the most remarkable sorties, passing invaluable information resulting in major changes in the original plans and speeding up of operations in this sector. A typical day's routine noted in the flight diary reads like this: *"Christmas Day – busiest – Flight Commander took CRA to the new strip .... While the CRA was busy in his work, flew a casualty, then rescued a pilot from a crashed Hurricane in no-man's-land to the Flight strip .... Captain Day flew a contact sortie .....Captain Hadley flew down to 74 Brigade on the coast with 'hot' operation orders.... Captain Hutt flew 51 Brigade Commander on a contact sortie over Pt 1296 .... Captain Boyd flew numerous contract sorties for 74 Brigade and shot 8 Medium Regiment on to Pagoda Hill .... Captain McLinden looked for three guns marked on a captured Jap sketch map ..... Registered two targets .... Captain Day set off searching for the second Hurricane pilot who had crash-landed whom he eventually found ..... over seventeen hours flying time in twenty one sorties."*

The final Arakan offensive started on 14 December 1944. Capt Boyd was attached to 74 Indian Brigade which was ordered to move down the coast. The coastal strip was narrow but the beach was wide and firm, spanning the whole length of the coast, offering a perfect landing ground to the Auster. But the problem was the sand which would get into the oil and petrol systems. On 28 December, Capt Boyd had an engine failure and force-landed five miles inside the supposed Jap territory. Making a lonely and hurried hike for the hills beyond which were own troops, he came across a small boy near a village who fled, yelling something. But then the whole village turned out and to his relief told him that the Japs had left. Capt Boyd now was leading a procession of Burmese, gathering numbers as they went and was first spotted by an Air Sea Rescue Hurricane which was looking for him and then was met by a Bren Carrier patrol that was sent to find him. But the Auster could not be retrieved as by the time ground party got there, it had been stripped by the locals who showed the ground party their new clothes made from the aircraft fabric.

With R/T communication always a problem for the widely stretched ground formations, the flight achieved the astonishing feat of keeping in touch with Squadron Headquarters 250 miles away, separated by the 10,000 ft high mountain range. It was over the squadron R/T on the second day that Maj Coyle received the message from Gen Sir Oliver Leese Commanding ALFSEA: *"I want to congratulate you on the magnificent work you fellows are doing on all fronts. Everywhere I go, I hear praise from everyone on the work the Air OP is putting in."*

Capt Hutt of 'C' Flight enjoyed working with 81 West African Division who were on a complete 'man-pack' basis moving down the trackless Kaladan Valley as the Africans were experts at making strips as a matter of routine. If halted for 24 hours, 1,000 Africans, equipped with their wicked-looking machetes would turn out on to the paddy field and a strip would be ready in a matter of a few hours. The ground parties had to walk with the ground troops from landing ground to landing ground, with the Africans carrying their tool kits on their heads.

Now let us come to the remarkable story and the notable part played by Air OP pilots in the final phases of the Burma campaign i.e. capture of Akyab on New Year's Day 1945. This was planned as a sea-borne operation which entailed an extensive bombardment of the island and the beach defences by an imposing force of one 15-inch gun battleship, three 6-inch gun cruisers, six destroyers, 48 Mitchell bombers, 72 Thunderbolts, 36 Hurribombers, 48 Beaufighters, 12 Lightnings, 12 Librators, and two dozen Spitfires as well as a regiment each of field, medium and heavy anti-aircraft guns and two 7.2 inch Howitzers – all this to be observed by Air OP.

On D-1, when two registration shoots had been done, Capt PJ McLinden was asked to fly a reconnaissance sortie to check the rumours that the Japanese had vacated the island. Before take-off, four heavy and three light anti-aircraft positions, along with several anti-aircraft machine gun areas and also the defensive lay-out of the Japanese, as per intelligence reports, were indicated to him during his pre-flight briefing.

He flew over the beach defences first at 3,000 ft and seeing no signs of occupation or enemy movements, flew over the town and the whole of the island, gradually descending to 1,000 ft and finally to tree-top level. The inhabitants waving white flags came out in large numbers, imploring him to land on the main airstrip. He then flew low over the extensive defences and found them abandoned and dilapidated. When the pilot returned with all this information, Royal Artillery Headquarters passed this information to the Division and Corps Headquarters asking for a stop to the operation, at least, the bombardment of the island to avoid casualties to the civilan population which was known to be pro-British.

The first reaction was that the operation must proceed, as "it would be a good chance to practice the 'Combined Operation' of the three Services" as months of hard work had gone into its planning. But to make sure that there were no local inhabitants in the target area, Capt Jarrett was asked to fly another sortie.

Jarrett saw that the locals were busy cutting paddy and making hay stacks – all waving white flags. He also found, outside the deserted town and the village to which all the population had moved, a huge crowd and a perfect ALG, made overnight, each corner marked by four men with white flags. He decided to land there and was mobbed by a cheering crowd of civilians, some talking English and shouting that the Japs had gone (the last 15 the night before). They requested him "to come quickly" as dacoits were terrifying them and also not to bomb or shell. Jarrett, ploughing his way through the crowd, took off and reported the story to GOC 25 Division, who feared that it was too late to do anything. But then, almost a crusade to stop the operation was waged, led by the Commander Royal Artillery and his Commanding Officers who said that they were going to drop all their shells in the sea. Just at that time, the Corps Commander arrived and was told the story. He was somewhat bewildered to hear it and said that the Supreme Commander had ordered the operation to go ahead but he would think about it. Better sense prevailed and to the relief of everyone, the orders came that the landing exercise would go through but without the bombardment. At the same time, Jarrett was asked to go and fetch the headman, if he would come. Taking Gunner Carter with him, he landed at Akyab again amongst the cheering crowd. When he asked for the headman about ten 'headmen' tried to get into the Auster. Promising to come back, he took one, leaving Gunner Carter behind but with a large hole in the tail plane caused by someone who stood on it for a better view. Carter who was treated like a 'king' overnight, served with coffee and fried chicken by the elders, could be called the first 'Military Governor of the Island'.

On D-Day at H minus 1, Capt Boyd flew to three cruisers which could not be contacted by wireless and told them over the R/T that the shooting was off. Later, throughout the day, Air OP pilots flew the RAF photographer, commencing at H-hour from the landing of the Special Service Brigade till the roll-in of the complete invasion fleet.

After securing Akyab, the 'island hopping' began, covering five islands, before landing in Rangoon. All these landings were covered by the pilots of 'C' Flight, with the Austers fitted with additional long range tanks. Flying over the sea and beaches and keeping stations in contact with cruisers and destroyers besides the ground troops was an altogether different experience for the Air OP pilots. Capt Boyd, while covering the landing on Myebon peninsula, was flying around the HMIS *Nabada,* trying to net his wireless set when he was shaken up by two large splashes straddling the ship. Looking up, he saw eight Japanese 'Oscars' on a dive bombing raid on the *Nabada* and the ship putting up a heavy barrage of AA fire against the raiders. He also saw an 'Oscar' diving down towards him. With a hostile coastline and own troops not yet ashore, he wondered where he could carry out the advocated 'evasive low flying'. But the 'Oscar' diving at him splashed into the sea 400 yards ahead of him, having been hit by the AA fire. Boyd's narrow escape was narrated by his friends on the ship thus "Wally Boyd flying tight circles at wave level round and round inside a small cove with his eyes shut tight."

The pilots of 'A' Flight, in support of 33 Corps, as mentioned earlier, now facing the Irrawaddy crossing in Central Burma, were also flying to their limits. The month was January 1945. The type of support Air OP provided in a large number of sorties flown everyday is highlighted by two citations, first the award of an immediate Military Cross to Capt Maslen-Jones supporting 20 Indian Division and the DFC to Capt Gregg:

> Capt Maslen-Jones was Artillery Air OP Observer to 32 Ind. Brigade Group during the Brigade's operations to capture Monywa. The enemy was well dug-in in perfectly camouflaged positions which proved most difficult and expensive in casualties for infantry ground patrols to locate. Capt Maslen-

Jones, in addition to his normal duties, and at great personal and continuous risk from enemy small arms fire, made many tree-top reconnaissances over enemy positions in compiling sketches on which were based the final plans for artillery-air strikes for the assault on Monywa. It was only after the capture of Monywa and after examination of the enemy positions that the full value of the daring work done by this officer and of the risks he must have taken in doing it could be appreciated. It is no exaggeration to say that the success of the assault on Monywa and the comparatively small casualties incurred by our infantry are due in no small measure to the initiative, continuous gallantry and devotion to duty with which this intrepid young officer carried out his reconnaissance.

Capt Gregg's citation read as follows:

On the 30 January, 1945, at Thayabaung, he carefully examined the enemy positions, and drew a sketch in the air of their defences. This he dropped on the Battalion Headquarters. It showed that existing maps were completely out of date, and was the means by which the successful attack was staged. He later dropped up-to-date maps into the very laps of the FOOs and the attacking company commanders, further ensuring success when other means of communication were uncertain. In this operation, his direction of artillery fire was outstanding, as the numbers of Jap bodies found later in areas engaged by him testified. Again, on 4 February, 1945, he was in the air directing medium artillery fire into Jap positions at Nyaungyin at such an early hour before the possible arrival of ground OPs and infantry that he attained surprise and averted a dangerous situation by causing the enemy to evacuate positions before they had time to dig in. His keenness to go into action and extreme willingness to do any task allotted to him have been a stirring example to all ranks.

During this most busiest and crucial phase, 'A' Flight was reinforced by two pilots who were sent to relieve the pressure of the overworked detached sections; they were Capt Rogers and the first Indian officer Capt FSB Mehta (nick named Duck Mehta)[1] who were trained at Jodhpur and Ranchi. In the last two weeks of January and in the four weeks of February, the six pilots of 'A' Flight flew 600 hours in over 600 sorties in six Austers.

Now let us turn to 'B' Flight which went into action in the second half of February 1945. The flight was actively engaged in supporting 17 Division in its dash to Meiktela. Since the town lay across the vital lines of communication to Mandalay, the Japanese 'reacted' violently and counter-attacked when 'B' Flight had to abandon their first strip as it had come under harassing fire by the Jap artillery. Luckily, all the aircraft had been dug in and the only damage was from the odd splinter. The flight managed to evacuate all the aircraft. During the Meiktela battle, the 'B' Flight pilots located 14 enemy guns, the largest number at one time for this campaign. Throughout April, the flight was always involved in the vanguard action in the advance by 33 Corps to Rangoon which was entered at the beginning of May 1945.

The highly successful work of the widely separated flights would have been impossible without an equally strong back-up support by the Squadron HQ under the able command of Maj Coyle who always moved with the advanced elements of the 14 Army HQ. Besides obtaining supplies of aircraft and spares from the depots as far back as Calcutta and Ranchi, he kept himself updated about

---

1. Capt FSB Mehta (later Brigadier) was the only Indian Air OP officer to serve in World War II. After the war, he was first detailed for the Staff College course at Quetta and then a Land/Air Warfare course in the UK. On return, after a short stint in the School of Artillery, he took over 9 Para Field Regiment in 1947 (the youngest CO at 27). He was too senior to be inducted in Air OP when No. 1 Air OP Flight was raised in Lahore on 15 August 1947. Always a hardcore Air OP officer at heart, he pushed the case and cause of Air OP throughout his service and besides being an inspiration and role model, was instrumental in the expansion of Air OP, especially when he was Brigadier Artillery HQ Western Command in the late Sixties.

the Army Commander's future intentions through the Brigadier Royal Artillery 14 Army under whom the Squadron was operating. The Squadron HQ Workshop and Equipment Section, each under RAF warrant officers, did a splendid job to keep the flights in fighting trim. The Austers were flown out from the flight for major overhauls but invariably the replacement aircraft reached first and the flights never felt the shortage of aircraft. The Auster was the method of transport which carried everything from a split pin to a spare engine. Maj Coyle often flew to the flights for any trouble shooting. The Squadron HQ was normally not authorised to hold aircraft but Maj Coyle always had two or three Austers and a reserve of pilots to undertake photographic, communication and supply sorties to the flights.

As mentioned earlier, the pilots encountered conditions some of which could be described as the most inhospitable in the world. Like a strip of 300 yards was made by bulldozing the top of a mountain spur at 6,000 ft with a sheer drop of 3,000 ft at each end, which required a precision approach to land and stop the Auster within 300 yards but the take-off was always spectacular as the aircraft would first sink out of sight and take at least a minute of hard climbing to come to the level of the strip. Some of the communication trips were also very dangerous and were undertaken with very low safety margins. To reach C Flight on the Arakan front, a mountain range, 10,000 ft high, had to be crossed when a well-laden Auster had to 'struggle' to 'negotiate' anything above 7,000 ft. In April/May, flying conditions would further deteriorate, with the sudden onset of violent rainstorms. Once a pilot faced a wind speed of 80 miles per hour in which he was almost helpless and static. Another pilot tried to dive under a storm at 140 miles per hour (144 miles being the maximum permissible speed) but lost no height at all.

With the capture of Rangoon, the campaign in Burma came to an end. On 12 May, the whole of 656 Squadron had arrived in Rangoon. In early June, it left for India for a refit in order to prepare for the Malayan invasion but was sent to help in the fighting in Indonesia via Malaya. A most notable and highly commendable feature of Air OP operations in Southeast Asia was that there was not a single fatal casualty of Air OP pilots – a proud record indeed – in spite of the most harsh and challenging environment.

After-Action Reports

There was all round appreciation of Air OP pilots by the formation commanders. Some of their reports read:

> ..... The cub planes have been the answer to the artillery man's prayer ...... they have performed about 40 percent of our observed missions and more than 60 percent of our precision registration.
> .... In the hedgerow country (Normandy) where ground Ops were non-existent, the cub was the only thing that was effective against Jerry S P guns.
> .... A liaison pilot of the 87 Infantry Division has been credited with sinking three ships and two barges on the Rhine.
> .... you wanted to know about your planes in action, well they are honeys! We call them our "air support" – and they're just that. Once they are up in the air, Jerry keeps his head down and his artillery stays pretty quiet. Ack-Ack and ground fire doesn't seem to bother them. I have seen them hightailing away from MEs and FEs (German fighters) and shake them too! They are doing an excellent job and the boys who pilot those planes deserve plenty of credit.
> .... Junction with Russians was most happily affected, first by personnel of the 10 Infantry Division Artillery in its cub liaison planes, next by ground patrols and then in an exchange of visits by staff officers and commanders.

Brig Gen Henry C Evans Artillery Commander 76 Infantry Division at the end of hostilities stated that over 70 percent of artillery missions fired and tactical information received came from liaison planes.

On one occasion in the Pacific, about 150 sorties were flown by liaison pilots to drop 'K' rations, ammunition, medical supplies, blankets and socks to the isolated troops on an island.

Often, the liaison planes were called upon to guide the fighters and bombers onto the target.

"Germany can't win," a Nazi prisoner told Lt Col John W Mayo.

Asked why, the German pointed to a cub circling overhead, and said, "Because of those things. That's the most dangerous weapon you have. We can't get out of our foxholes without being spotted by one and when our artillery shoots, they always spot it and bring down counter battery."

When Col Mayo asked why the Germans didn't shoot the cubs down, the prisoner replied, "Our officers have told us that the planes are very heavily armoured and couldn't be shot down." But the real reason was the fear of the inevitable counter battery fire.

An extract from a report submitted by a Panzer Grenadier Battalion Commander read:

> As a rule, an attack is preceded by a strong artillery preparation in which the Americans employ all calibres, including their heaviest. Planes are used for fire directions and excellent result has been obtained.... Artillery directed by observation planes places fire on each of our movements .... Whenever possible, attack preparation should be avoided during the day. US air observation detects every movement and directs sudden and heavy fire concentrations on the deployment areas.

Another prisoner, marvelling at the effectiveness of the Army's Air OP/Liaison planes, said that every time he saw an observation plane, his blood would boil. "It is an insult to have that little defenceless box bobbing around in the air and not be able to do something about it," he said.

Still another German prisoner stated, "When the cub flies over, all firing ceases. All we move is our eye-balls."

A Japanese prisoner said that more fear was generated by the sight of a cub above them than by any of our other planes. The reason was that, invariably, when they saw cubs, artillery was brought down on them and more of their men were killed by this fire than by bombs.

German respect for the liaison planes was also revealed in a captured document which established a point system for awarding German fighter pilots a medal. Three points were given for knocking down an escorted four-engine bomber, two for a two-engine escorted bomber, one for a fighter plan, and two for a liaison (Air OP) plane.

Casualties, Honours and Awards – British Army

Commencing with 'D' Air OP Flight, formed on 1 February 1940, a total of twelve Air OP Squadrons each consisting of three Air OP Flights, were raised between 1 August 1941 to 30 September 1943 in the British Army. In all, 520 Air OP officers from the Royal Artillery, 73 officers of the Royal Canadian Artillery, 2 from South Africa and 1 from the Royal Dutch Forces were trained upto VJ Day 1945 at Larkhill. A small number serving in the Royal Artillery units in India, including one Indian officer, Capt F S B Mehta, were trained in India.

The casualties suffered and honours and awards won by British Air OP officers during this war are given below:

Fatal casualties.
Operational	37
Non-Operational	24
Total	**61**

Breakdown of operational casualties
North Africa	01
Italy	18
NW Europe	18
SE Asia	Nil
Total	**37**

Main causes of casualties
Anti-aircraft fire	03
Enemy aircraft	06
Small arms from ground	04
Own shells	09
Flying accidents	08
Unknown	07
Total	**37**

Honours and Awards (British Air OP units)
DFC	101
DSO	02
MC	06
MBE	11
Polish Cross of Valour	03
Croix de Guerre (French)	09

4

AIR OBSERVERS IN THE US ARMY

Bad reconnaissance had led to spectacular military failures..... Good reconnaissance gives a commander a regular, small advantage and, in battle, it is the sum of these small advantages that gives victory."

— Brigadier Peter Mead—A World War II
Air OP pilot and first Director,
Army Air Corps, British Army

Upto World War I

To complete the history of Air Observation, let us connect with similar events which took place before, during and after World War I in the United States Army.

Although the Balloon Corps was disbanded in June 1863, Army Commanders did not forget the value of aerial observation and the contribution made by Lowe to the concept of Army Aviation. Recognising the significance of Lowe's contribution, much later, the Army honoured Lowe, naming the airfield at its primary fixed wing training establishment at Fort Rucker, Alabama, as the Lowe Army Airfield.

The concept of aerial observation was resurrected with the reestablishment of the Balloon Corps as part of the Signal Corps during the Spanish-American War. This was due to the foresight of two officers, Brig Gen Adolphus W Greely and Brig Gen James

Allen. Gen Greely was the Chief Signal Officer from 3 March 1887 till 9 February 1906 when he moved out on transfer and promotion and was replaced by Gen Allen who continued with the 'aeronautic policies' of his former chief. In 1892, the Signal Corps plans called for a balloon section as part of each 'telegraph train'. A balloon was obtained from a French firm and named 'General Myer' in honour of the first Chief of the Signal Corps. This balloon was used extensively over the next few years until destroyed by high winds.

On the outbreak of the Spanish American War in 1898, the Army had only one balloon. It was named 'Santiago' as it was planned to be sent to Cuba but was instead moved to New York to watch for any anticipated invasion of Manhattan by the Spanish. But realising that such a bold attack would never materialise, Gen Greely suggested the balloon to be sent to Cuba where, after numerous transportation snarls, it finally arrived on 28 June 1898. Three ascents were made in the balloon on the afternoon of 30 June at the request of the commanding General, Maj Gen William Shafter. The observations provided the Army with valuable information on the roads to the front and the location of the Spanish fleet in Santiago harbour.

On the morning of 1 July, the Army prepared to launch an attack on San Juan Hill and to destroy a heavily garrisoned place called Blockhouse, the last remaining obstacle on the road to Santiago. Col Joseph E Maxfield, in charge of the Balloon Corps, was ordered to have his balloon keep pace with the lead units. Winding its way among the advancing troops, the balloon was soon within a few hundred yards of the Aguadores river where it came under heavy fire from the Spanish Army's field guns and musketry fire. It was hit and torn badly. It descended but afforded enough time to Col Derby who had gone up in the balloon to discover a road leading from the main road to the left and crossing the Aguadores river, four or five hundred yards farther down the stream. This was the most important and timely information as the main road was congested with troops, with heavy enemy fire, which could have resulted in very heavy casualties. Gen Greely later remarked, "The action enabled the deployment of our troops over two roads and by doubling the force, may possibly have been the determining factor in the gallant capture of San Juan Hill."

After the Spanish-American War, the Army's balloon activity remained fairly stagnant till June 1907 when the Signal Corps purchased their 9 and 10 balloons. It was the vision of Army officers like Gen Greely who knew the value of aerial observation that kept the concept alive in spite of numerous setbacks, including opposition within the Army. Gen Greely's good work was continued by Gen Allen who took over as Chief Signal Officer from Greely on 9 February 1906.

Under the guidance of Gen Allen, a balloon house and hydrogen plant were established at Fort Omaha in 1908 but the ballooning in the US Army retrogressed over the next few years. When the United States entered World War I, the Army had only three serviceable free balloons and two captive balloons on hand. A crash training programme was undertaken at Fort Omaha and new balloon companies were sent to field artillery firing centres. Balloon observers, after receiving training at the American schools, were sent to French schools and artillery centres. By Armistice Day, the Army had 89 balloon companies and 751 balloon officers in America. Thirty-three companies and 117 officers were sent overseas to join 2 balloon companies already organised in France. Of the 265 balloons sent to France, 77 participated in the war. The US Army employed 252 balloon observers within the balloon companies. Specific tasks given to the balloon observers were: recording heavy artillery fire, shot by shot, observing demolitions behind enemy lines and watching for reinforcements or traps, shifting of enemy positions, assembly of supplies by the enemy and the forward movement of enemy troops. The Army lost 48 balloons; official German losses were set at 73. In all, the Army's balloon operations in World War I accounted for 1,642 ascensions (3,111 hours in the air) 316 artillery adjustments,

12,018 shell bursts reported, besides numerous other types of information given to the ground troops.

After the historical flight by the Wright Brothers on 17 December 1903 which was followed by further experiments when the heavier than air flying machine became a reality, the United States became the first country to contract for a military aeroplane. An Aeronautical Division was created on 1 August 1907 as part of the Signal Corps at the direction of Brig Gen James Allen. The Aeronautical Division which was made responsible for all matters pertaining to military ballooning, air machines and all related subjects, issued an advertisement asking for bids for a heavier than air flying machine, laying down the performance parameters.

On 1 February 1908, the Army received 41 bids but only three bidders met the requirements outlined in the specification. All the three bids were accepted but only the Wright Brothers who had quoted $ 25,000 to make the machine in 200 days delivered the aeroplane. The machine was brought to Fort Myer, Virginia, for trials which commenced on 3 September 1908.

On 17 September 1908, Orville Wright invited Lt Thomas E Selfridge to ride as a passenger in the third test flight. Selfridge had earlier made a solo flight in a powered airplane on 19 May 1908 and was considered the most widely informed expert on the dynamics of the air and mechanical flights and was also nominated the official Army observer for these trials. Just as success seemed imminent, tragedy struck when, during the fourth turn, one of the props struck a brace wire attached to the rudder and the machine twisting and turning fell 150 ft. Selfridge died a few hours later but Oriville survived the crash and remained hospitalised for several weeks. Upon his discharge, he and his brother continued their work and returned to Fort Myer on 20 June 1909 with an improved version of the machine.

After a series of practice flights, Orville again announced readiness for the official trials. On 27 July 1909, he made the first test flight carrying Lt Frank P Lahm as a passenger. Lt Benjamin

D Fonlois flew with Orville on the final test on 30 July 1909. The tests were successful and the Army accepted the aeroplane on 2 August which became the first US Army Aeroplane. Orville trained three officers, Lt Humphreys, Lt Lahm and Lt Fonlois to fly the aeroplane as part of the contract.

Due to non-availability of funds, the Army struggled along with one pilot (other pilots returning to their regimental duties) until 1911 when Congress appropriated $ 1,25,000 for Army Aviation. Gen Allen received $ 25,000 immediately and placed orders for five planes. By November 1912, the Army had 12 pilots, 39 enlisted men and 12 airoplanes in its inventory.

The Army first used aeroplanes for observation and adjustment of field artillery in November 1912 when, at the request of the Field Artillery Board, two aircraft were sent to Fort Riley, Kansas, for a series of experiments.

In the first test on 4 November 1912, flying was done to establish the type of signals to be used between the aeroplane and the battery. For this purpose, although not very reliable, a wireless station was put up in the immediate vicinity of the battery. But a card system where the observer plotted the position of the hits on the card in relation to the target and dropped it as he passed over the battery, was found quite workable and effective. In the second test on 5 November 1912, actual firing was done on a target which was about 3,400 yards but invisible from the battery. The target was easily picked up by the air observer and was hit after four 'volleys'. In the subsequent trials, more targets, invisible from the battery, were engaged with greater success, achieving a hit at the third volley.

In August 1913, a Bill in Congress called for a separate Aeronautical Branch for the Army which was opposed by the Signal Corps Officers. However, later, on 18 July 1914, Congress created an Aviation Section within the Signal Corps. The Aviation Section increased Army Aviation's strength and scope, gave it a definite status and provided manufacturers much needed encouragement.

During World War I, the Army had 39 aero squadrons which participated in action. These included 12 Corps Observation, 3 Army Observation, 18 Pursuit, 1 Night Bomber and 5 Reconnaissance. Assigned to these units were 1,402 pilots and 769 aeroplane observers.

Corps and Army Observation Squadrons were responsible to meet the demand of 'aerial artillery observation'. When a mission was requested, an aircraft from these squadrons was dispatched which first established the radio contact with the artillery unit, got detailed briefing about the task, carried out the task and then returned to the base to await another assignment.

Debates and Trials
With the signing of the Armistice in November 1918, and return home of US military personnel came the men who had new ideas about waging war through air power. New and better aircraft were developed and the concepts of strategic bombing and air superiority grew. Extremists, as in the UK, advocated that combat air power used on a mass scale could break the enemy's will to resist and only minor operations from the ground forces would be needed to win the war. Amongst these heated and sometimes passionate arguments, developed the framework of the Army – the concept of Army Aviation.

In the meanwhile, the War Department had first created, on 24 May 1918, a single agency, the Air Service, to take care of all operations, including aircraft production and military aeronautics which was followed by the creation of the Air Corps by Congress through the Air Corps Act of 2 July 1926. This Act also created the position of Assistant Secretary of War for Air.

Aware of the support that balloons and aeroplanes had provided in World War I, the Chief of the Artillery stated that air observation was vital to the effective employment of artillery and ordered a thorough study on the artillery's experience in combat with aerial observation.

Disagreement between the artillery and the Air Corps grew as the Air Corps was rapidly developing its concepts of strategic air power, weakening its entire ground support programme. The Air Corps also felt that merely to furnish an aeroplane and a pilot to carry an expert observer would be to relegate the Air Corps to a status more of a service than a combat arm.

As the 1930s drew to a close, artillery officers fired an increasing 'barrage' of demands for more effective aerial direction of fire. When these demands were not satisfied by standard Air Corps Observation Squadrons, artillery officers began advocating the use of light aircraft organic to the their artillery battalions.

In the summer of 1940, Lt James McChord Watson, an artillery officer, directly contacted the Piper Aircraft Corporation and discussed the artillery's plan to use the light aircraft for adjustment of fire and arranged for an aircraft to participate in manoeuvres to be conducted in August 1940 at Camp Beauregard, Louisiana. On 12 August 1940, Mr Tom Case of Piper flew a light aircraft to the exercise area and with Watson as observer, took part in the manoeuvres after which they remained in touch to discuss and find solutions to many connected problems.

Interest in the light aircraft was also mounting throughout the Army. Gen Adna R Chaffe at the Armour School, Fort Knox, Kentucky, was intensely interested in using the light aircraft to control armoured columns and adjust heavy cannon fire from tanks. Mr Case from Piper flew to Fort Knox on 10 February 1941 and carried out evaluation flights in the Armour School till 15 February. At the request of Gen Horace Whittaker, Commanding General of 45 Infantry Division; Mr. Case also flew to Camp Bowie, Texas, and carried out numerous demonstration flights between 17-23 March 1941. Gen Whittakar also began corresponding on the subject with the Chief of Field Artillery, Maj Gen Robert M Danford, a dedicated advocate of making light aircraft organic to the field artillery.

On 18 February 1941, Mr. William T. Piper, Senior President of Piper Aircraft, wrote a detailed letter to the Secretary of War pointing out the great potential of light aircraft in support of the ground forces. Mr. John E P Morgan, who was the sales representative for Piper, Aeronca and Taylorcraft, and was based in Washington, mounted pressure on the War Department for an expression of policy on the employment of light aircraft so that the industry could consolidate its efforts to provide the most efficient cooperation and the best aircraft.

In May 1941, another event gave further impetus to the case of aerial observation of artillery fire – Maj William W Ford, a field artilleryman, aviation enthusiast, pilot and sportsman who had been working intensively to bring organic aviation into field artillery, wrote an article (which was considered quite explosive) outlining his concept. The article impressed Gen Danford and was published in the *Field Artillery Journal* in May 1941 (Maj Ford was later destined to direct field tests and became the first Director of the Department of Air Training at Fort Sill, Oklahoma).

Meanwhile, the Piper Aircraft Corporation, along with Aeronca Aircraft Corporation, Taylorcraft Aviation and Continental Engine Company formed a team which later came to be known as the Grasshopper Squadron behind which there is an interesting story. Flown by their civilian employee pilots, these companies started offering their light aircraft for trials to the Army at their own expense. Their sales representatives also started contacting Army Commanders directly. In one such contact, Mr Henry Swann (later Lt Col), district sales manager for Piper Aircraft in the Western States, was put in touch with a Colonel at Fort Lewis who told Mr Wann that he knew light aircraft had a great deal of potential, especially for artillery fire adjustment as he himself had a private licence and was well aware of the light aircraft's uses. When Wann asked his name, he said, "Eisenhower." Later, they met during the Third Army manoeuvres between 1-30 September

1941 in Louisiana where Col Eisenhower – the future President of the United States – was Chief of Staff.

In May 1941, Mr Case from Piper Aircraft flew a light aircraft to Fort Sill to take part in the army exercises. He was joined by more pilots later. They flew numerous missions with army officers in the second seat to observe troop movements and artillery fire. Later, from 2-28 June 1941, twelve aircraft participated in the Second Army manoeuvres in eastern Tennessee in which they competed with regular Air Corps Squadrons that had brought their own (faster) aeroplanes. Serving both the "Red" and "Blue" Armies, they were given similar tasks. Light aircraft operated from small clearings and roadsides from near the ground troops they supported whereas the Air Corps pilots operated from their rear bases. Tasks included adjustment of fire from 155 mm guns, 'scouting' advanced enemy positions, flying commanders and staff, reconnaissance and column control missions and directing cavalry operations. While the light aircraft completed all the missions successfully, pilots from the Air Corps Squadrons were not able to identify the targets most of the times and aborted many reconnaissance missions.

During desert manoeuvres in July 1941 in Texas, which culminated on 26 July, Mr Wann was told to proceed to Headquarters, 1 Cavalry Brigade, from Fort Bliss, to deliver a message to the Brigade Commander, Maj Gen Innis P Swift and then remain with the General, till relieved. Mr Wann flew to the area and landed in a clearing but not before bouncing a few times on the rough ground which had some grass clumps. Taxiing to the Command Post, he delivered the message and informed the General that he had been instructed to remain with him for use as desired.

Gen Swift, quite impressed, invited Mr Wann to lunch but remarked, "You looked just like a damn grasshopper when you landed that thing out there in those boon-docks and bounced around."

During lunch, a trooper marched in, saluted and handed over a radio message just received, to Gen Swift that an aeroplane had just been dispatched to him for his use. It had taken the message 45 minutes longer to arrive than the aeroplane. The General used the light aeroplane continuously during the rest of the manoeuvre and named the aircraft "grasshopper." The tag stuck and this team of 12 aircraft which took part in these manoeuvres came to be known as the Grasshopper Squadron.

In the summer of 1941, Maj Gen Robert Danford, Chief of Artillery, US Army, who was already enthused about the performance of the light aircraft in the Army manoeuvres, visited the British School of Artillery, Larkhill, where he was given detailed briefing on the Air OP which, after the initial setback during the whirlwind German offensive, was being resurrected, as mentioned earlier. On his return he obtained permission to evaluate the planes in subsequent Army manoeuvres in Louisiana between August to November 1941. Twelve light aircraft from the Grasshopper Squadron first flew 12-14 hours a day in the Third Army manoeuvres from 11-30 August and then in the combined Second and Third Army manoeuvres from 1-30 September. Later, members of the Grasshopper Squadron continued to support the First Army manoeuvres held between 6 October to 30 November 1941.

During these manoeuvres Col Eisenhower, Chief of Staff, Third Army, and his staff constantly used these aircraft to observe the manoeuvres (the cavalry-tank skirmishes in particular) from the air. Gen George Patton was also a frequent user of these planes. He also had his own plane at the manoeuvres and often directed armoured columns from the air. Air Force observation units provided by the Air Support Commands also participated in these manoeuvres but failed to meet the demands of the ground forces, whereas the Grasshopper Squadron proved its worth beyond doubt and fully met the requirements of the Army units they served. This was forcefully substantiated by Gen Danford: *"The only uniformly satisfactory report of air observation during the recent maneuvers comes from those artillery units where light commercial planes operated by civilian pilots were used."*

The light planes' participation in these manoeuvres was considered a complete success and resulted in a recommendation to the War Department that light aeroplanes be made a regular component of the artillery. The Army requested permission to purchase 20 such planes but the request was disapproved by the War Department. However, Col Dwight D. Eisenhower arranged to have the pilots and the planes placed on a *per diem* rental and expense basis. Before that, members, of the Grasshopper Squadron paid their own way.

Subsequent to these manoeuvres Gen Danford recommended to the War Department that light aircraft manned by artillery officers be made organic to the division and corps artillery units. *He denounced any system which did not make light aircraft organic to his units. He was not prepared to accept any outsider acting as observer since an artillery commander would not know the observer and this was considered a critical point as artillery felt that "the point of fall of the artillery shell is an inextricable element of command. The artillery man cannot do his job if he surrenders this element of command to some stranger who is not responsible to him, whom he never sees and, therefore, who he cannot trust."* The Division and Corps Commanders unanimously endorsed this recommendation. Unfortunately, Maj Gen Leslie J McNair, Chief of Staff at GHQ, who generally favoured the massing of support elements, disapproved the recommendation, with a rider that a fair trial be given to the new system. As a follow-up, he ordered a trial of Gen Ford's proposal with a corps artillery brigade and an infantry division. Lt Col Ford was placed in charge of the trials and ordered to proceed to Fort Sill, Oklahoma. A proper trial directive was issued based on which a full trial programme, to include selection of personnel, including artillery officers, to be trained as pilots and first echelon maintenance of the aircraft was issued by HQ Field Artillery School, Fort Sill.

This was the time when the United States was mobilising for a full scale war after the Japanese attack on Pearl Harbour on 7 December 1941.

2 Infantry Division and 13 Field Artillery Brigade were nominated as the trial units for the field trials commencing 15 January 1942. The trials which included basic and tactical flying training, landing on roads and improvised strips, field handling and procedures for adjustment of fire were completed by the end of April 1942.

The success of the trials was reflected in two separate letters of commendation by Brig Gen John B Anderson, Commanding General, 2 Infantry Division and from Brig Gen J A Crane Commander, 13 Field Artillery Brigade. They admired the skill, enthusiasm, outstanding work and the undertaking of seemingly impossible tasks, free of serious accidents, with total disregard of personal inconvenience by the trial team.

On 5 May 1942, Gen Crane wrote to the Commandant Field Artillery School that if the trial report was approved and air observation made an organic part of the field artillery, the credit would go to the trial team, and the Field Artillery arm of the Services would owe them a "debt of gratitude."

The Boards appointed to observe the test forwarded their reports to the War Department, highly recommending organic aviation for field artillery units. GHQ Army Ground Forces also was impressed and recommended that the programme be implemented without delay. Gen McNair, who earlier was sceptical of this concept, was also convinced of its worth and "supported it with all his power."

Ultimately, on 6 June 1942, the War Department issued a Memorandum – Subject: *Organic Air Observation for Field Artillery* – approving the Army Aviation Organic to the Army Ground Forces. The directive authorised two light aircraft with two pilots for each Field Artillery Battalion, two per Field Artillery Brigade or Group Headquarters and two per Division

Artillery Headquarters and one pilot per aircraft. It also contained instructions for procurement and maintenance, qualifications for pilots and mechanics and their training.

In anticipation of this decision, Maj Gen Robert M Danford had already done lot of preliminary work by even keeping the first batch ready for training, which commenced without any loss of time.

The first flying instructors consisted of civilian pilots, mostly the members of the test group. Eventually, most of these civilian instructors were commissioned and put through the tactical flight courses and special capsules were conducted to impart training in air observation and direction of artillery fire. Pilots were also given training in repair and maintenance of aircraft in the field and issued a kit of hand tools.

Army aircraft mechanics were responsible for all first and second echelon maintence in the field. Third and fourth echlon maintenance was the responsibility of the Army Air Force.

World War II

The Army pilots were referred to as liaison pilots and their aircraft as cubs in the US Army and were an integral part of the artillery battalions. They arrived to take part in the African campaign at the same time as the first British Air OP Squadron. Like in the British Air OP, battle procedures were evolved and refined under the actual battle conditions by the liaison pilots. No doubts, there were lots of lessons learnt from the mistakes/blunders, which is not uncommon in any war. Lessons were well assimilated by the time of the Normandy landing where 4 Infantry Division of the US Army was the first to take its pilots and aircraft across the English Channel. They faced the same problems as the British Air OP units. Some of the case histories and exploits are worth recounting.

Capt Gregorie, an artillery aviation officer from an artillery battalion of 4 Infantry Division, proceeded with the ground forces with most of the aircraft dismantled and loaded in trucks and carried across the English Channel to Normandy. A few aircraft were to be flown across on getting clearance from Capt Gregorie.

Arriving at Utah Beach on D-day, Capt Gregorie, at 1500 hours located the area pre-selected from the map for the landing ground but found that artillery fire had left it in an unusable condition. He searched for another area by next morning and radioed a message to Lt Dave Condon, his assistant, to fly along with the other pilots from England. Lt Condon who arrived at about 1500 hours, had some difficulty in locating the strip but with the help of smoke grenades and talking on the radio, Gregorie brought him on to the ALG. By this time, all the divisional artillery had been brought ashore and deployed but the batteries had not registered because of limited visibility from the ground. Capt Gregorie and Lt Condon immediately took off without refuelling and registered targets on Utah beach. This was the first registration by the liaison pilot of the US Army.

Maj J Elmore, Maj Swenson and Lt Harper, with their team of aviators from 29 Infantry Division and 1 Infantry Division, had also arrived at Omaha beach under similar circumstances and flew their first missions on D+1 day. All of them faced anti-aircraft fire but the cubs demonstrated the greatest operational efficiency against all odds and performed tasks beyond the expectations of the ground commanders on the Normandy battle front. They maintained excellent observation over the entire front except during the worst weather. Brig Gen George Shea, one of the Commanders in Normandy, stated: *"Without the cub plane, practically all our fires would have been unobserved. With the cubs, weather permitting, we could observe most of our fire. Any artillery man knows what that means."*

87 Infantry Division which was fighting the famous Battle of the Bulge, with most of the area around covered by snow, had the

problem of its cubs not being able to take to the air. This problem was overcome by getting skis for the cubs which was the first time that the cubs used skis in combat.

During the Battle of the Bulge, Lt Samuel Fein with 2 Armoured Division was also involved in a 'first'. While flying in a routine mission, he was startled by several German "buzz bombs" which whizzed past his plane. Circling, he located a portable ramp and saw another buzz bomb launched, the first time the Germans had launched these bombs. Lt Fein called in artillery fire from the American and British batteries and, in a few moments, the launching platform was destroyed. Thanks to the sharp eyes of the Army liaison pilot, a German *"first"* back-fired.

After break-out from the beach-head of Normandy, Allied armoured columns moved fast into the Brittany Peninsula. Liaison pilots flew as far as 40 miles from their ALGs. Often, they were the only source of contact with the rapidly advancing armoured columns. They flew road and bridge reconnaissance missions and passed the information on enemy positions to the advancing troops. Some units were reluctant to move unless a cub was above them to give information.

Like their counterparts in the British Air OP units, they displayed the same daring boldness, at times not called for. Like, during the liberation of Paris, Lt Ross Hazeltine and another pilot prematurely flew their aircraft, probably to be counted as 'first' and tried to land at the Longchamp Race Track on the west side of the city on 23 August 1944. Skimming the roof-tops over peaceful Versailles, passing near the Eiffel Tower and slipping under a bridge, they came over the race track and were shocked to see the biggest anti-aircraft gun ever seen by them. They saw a few Germans run into the gun pits, opening their machine guns. "We ducked for the river. They missed us but I'll never know how because we were flying at the fantastic speed of 60 mph."

During the rapid advance on the Rhine, the liaison pilots proved invaluable in controlling unit movements when ground communications were stretched and could not be used. On one mission, Lt Horace E. Watson of the 8 Armoured Division spotted a reconnaissance party held up by a blown bridge. He circled the area and found another bridge intact that the troops could use. He made a low pass over the troops, pulled back the throttle to idling and yelled, "follow me." They did, and in a short, while another town was liberated.

Brig Gen Edward T. Williams, Gen Patton's Artillery Chief, devised a novel plan to help support Gen Patton's assault on the Rhine river. After establishment of the beach-head, the Air OPs were to transport reinforcements and the tests indicated that 90 cubs could airlift an infantry battalion across the Rhine in about three hours. The aircraft were assembled but the assault across the Rhine was such a spectacular success that they were not needed.

Gen Patton was often up in the cub to observe and direct the movement of his armour columns. Once, he had a narrow escape, when suddenly attacked by a fighter who made three passes on his cub. On the third pass, the fighter pilot pulled out too low and crashed. Later, it was learnt that the attack was made by a Polish RAF pilot who had mistaken Gen Patton's cub for a German cub.

Recalling the incident, Patton laughingly admitted he had been frightened and related: "After the first pass, I decided I might as well take some pictures of my impending demise. There wasn't anything I could do, so I thought I might as well use the camera. But after it was all over, I found I had been so nervous I had forgotten to take the cover off the lens and all I got were blanks."

Cubs also directed supply convoys across Europe by reconnoitering the most effective routes to the front. Sometimes the pilots would land near a convoy to indicate the location of forward units. A number of times, they flew food, ammunition and medical supplies to patrols and isolated or trapped units. Near Luxemburg, pilots headed by Maj Carl I. Sodergren flew supplies to an isolated infantry battalion of 76 Infantry Division for three days which enabled the battalion to resist capture or destruction. Similarly, two infantry companies from the Ninth Army near Engelsdorf, Germany, besieged by superior German forces, were able to hold off until rescued, when Lt William and his team of cub pilots flew medical supplies and blood plasma and also directed Allied devastating artillery fire and barrages on the German forces.

On the Metz front, cub pilots supplied trapped units of 95 Infantry Division with medical aid, food, ammunition and dry clothing. Once Maj Elmer Blaha flew Maj Eugene Cleaber (an Army surgeon) into an area under heavy attack and then shuttled supplies in and evacuated the wounded from a pasture pockmarked with shell holes and battle debris.

On numerous occasions, liaison pilots landed in fields to rescue downed fighter/bomber pilots. A classic story of one such mission goes like this: a cub pilot once saw a fighter pilot, after crash landing in a clearing, run for the nearby woods. Considering it a routine rescue mission, the liaison pilot set his cub in the field and bumping along, taxied to the spot where the fighter pilot had disappeared. Soon enough, he saw a man running out of the woods and waving at him. To his shock and surprise, the man turned out to be a German soldier, waving a pistol. As the liaison pilot began taxiing for a quick take-off, the German caught up, shoved the pistol into the pilot's face but it didn't go off. The German pulled the liaison pilot from the cub. What followed was a wild scene, the pistol changing hands, the only 'weapon' to hit, both rolling and fighting on the ground. Finally, the decision went in favour of the American who got up wiping his brow, with the unconscious German lying below, when an Allied fighter pilot also walked up and said, "Say, that was quite a fight." The liaison pilot, bruised and still trying to catch his breath looked at the fighter pilot in disbelief and shouted: "You mean.... you saw all this...... and didn't come to help me?"

"Oh, you were doing a fine job," replied the fighter pilot.

Once Lt Ford, supporting 76 Infantry Division, spotted an enemy machine gun nest waiting to engage its advance column. He flew low, throttled back and shouted a warning but crashed as his cub touched the side of a hill during the turn. The cub was a total loss but Lt Ford survived, and fought alongside the infantry for a few days before returning to his section.

It was an SOP in the 3 Armoured Division for one artillery observer to remain in the air during daylight hours to observe enemy movements. By then, the Germans had ordered their troops not to fire on the cubs so as not to give away their positions, which prompted the liaison pilots often to fly on enemy lines.

During the fighting in Austria, a bridge over the Inn river which flowed into Danube, was proving particularly bothersome to Gen Patton because the Germans were rushing in troops and supplies over this bridge into the 20 Corps Sector. Gen Patton ordered the bridge to be destroyed and gave the task to the 11 Armoured Division. The bridge could only be observed by the liaison pilots. One battery of 155 mm guns and 27 rounds per gun were allotted for this task. Bets were made within the 11 Armoured Division on whether or not the mission could be accomplished. Capt Gregore and Lt James flew to the area but were immediately driven off by three German FW 109s. They returned later and again got down to the business.

The first ranging round landed well over the bridge, the second short and about 200 yards to the right and the third splashed under the bridge. Capt Gregore called back that the third round landed "50 short". He was informed that round four was on the way. It seemed that the round was taking forever to get there but suddenly there was an orange flash, followed by a huge, billowing cloud of black smoke which engulfed the bridge, with the ends of the bridge sliding down into the smoke and river. The bridge, as discovered later was mined and the fourth round was a direct hit on the mined portion of the bridge. Staring in silence for a while at the spectacle,

Capt Gregore and Lt James regretted that they couldn't think of an appropriate statement to fit the occasion. They just said, "Mission accomplished" and went back.

"Mission accomplished," in fact a seemingly common statement was becoming uncommonly significant as the Air Observation/liaison pilots were echoing it hundreds of times daily in all theatres of war after carrying out missions considered impossible a few years ago.

Ingenuity and initiative on the part of the liaison pilots also were evident in the South Pacific where the American forces were doing island hopping.

Gen Joseph W (Vinegar Jeo) Stilwell, Commander 10 Army, used the cubs extensively. He enjoyed flying in the cubs with both windows open. On one occasion, on a mission off Okinawa, being flown by S/Sgt Lyle W White, a gust of wind blew out the General's battered campaign hat that he had worn for over 20 years. Gen Stilwell sadly watched his hat float down into the sea and remarked, "I'd sure like to go down there and rescue my old friend." On landing, General offered $ 25 to anyone who recovered his hat. Much to his delight, it was returned to him four hours later.

In 1945, before the fighting came to an end, a company of the 14 Infantry Regiment was surrounded in North Luzon in the Pacific. First, the liaison pilots directed the transport plans in dropping food and ammunition to the surrounded troops and then guided tactical air strikes which enabled the company to break out and escape.

In another sector, Lt Hartwig and Lt Laroussine, while flying, saw something move near a funny looking stack in a coconut grove where, as discovered later, a group of Japanese was huddling to keep out of sight. They called in the artillery and the second round landed squarely on the target. They 'fired for effect' – actual count showed 82 enemy dead.

Beginning 1794, this ends the story of Unarmed Air Observers till the World War II.

5
AIR OBSERVATION IN THE INDIAN ARMY

> The Air OP had proved its value in this campaign (World War II)... it has become necessary part of Gunnery...
> — Field Marshal the Viscount Montgomery
> Dispatch 1 June 1946

> Experience is the name everyone gives to their mistakes.
> — Oscar Wilde

The Beginning

We have seen how the Air Observation Posts commencing from the balloon era when man took to the air for the first time, evolved using various platforms and became an essential and integral part of the fighting element of the land forces in all the small and big wars till the end of World War II. Thus, the Air OP became a forerunner to Army Aviation after World War II. The US Army had taken the first step to integrate the observation aircraft with their artillery battalions on 6 June 1942 and pushed for taking over more elements of air power under their control after World War II and succeeded, but other countries took longer to form their air corps as an independent Service.

France formed it in 1954, Germany in 1956, the UK in 1957, Pakistan in 1958 and Australia in 1968. As this is a story of Air

Observation, logically it should end with the battle accounts of Air OP pilots in World War II.

However, the Air Observation units in the Indian Army continued to function (as Air Force units) for the next 41 years, till the establishment of the Army Aviation Corps on 1 November 1986. Therefore, this story has been extended to include battle accounts of Air OP pilots in wars fought in the Indian sub-continent till 1 November 1986.

There was no Air OP Units raised for the British Indian Army during World War II. As discussed earlier, it was 656 Air OP Squadron RAF, raised on 31 December 1942 at Westley, UK, which supported the Allied Armies in Burma from January 1944 to May 1945. In June 1945, it was sent to Ranchi for a refit and then sent to Indonesia via Malaya.

659 Air OP Squadron RAF which was formed at Firbeck on 30 April 1943, after taking part in the European campaign, was shipped to India on 2 October 1945, arriving in Bombay in the last week of October. From there, its flights and aircraft were detached to Deolali for operational training of new Air OP pilots, mostly of local Indian descent and to the North-West Frontier, to support troops dealing with local hostilities and, subsequently, to Punjab to undertake reconnaissance of the disturbed areas on the outbreak of communal riots. The Squadron HQ was based in Peshawar. By June 1947, the Auster MK 5 of the squadron were deployed as under:

- Six aircraft at Peshawar with Sqn HQ;
- Three aircraft at Lahore;
- Four aircraft at Deolali;
- Two aircraft at Razmak; and
- Two aircraft at Jullundar.

However, after a month and a half, on 14/15 August when India got "Freedom at Midnight," the Squadron disbanded in theory and with its assets, two independent Air OP Flights, one for India and other for Pakistan, came into being on 15 August 1947. By then, a number of Indian officers from Royal Artillery units of the British Indian Army had undergone their flying training in the UK.

Maj H S Butalia, thus, raised No. 1 Air OP Flight at Lahore on 15 August 1947. On 13 September 1947, he, along with Capts Govind Singh, Sridhar Mansing and RN Sen flew to Rajasansi airfield near Amritsar.

Later, No. 2 Air OP Flight was raised at Deolali on 1 November 1947 with the assets left by 'A' Flight. After handing over the assets, the British personnel of 659 Air OP and its flights returned to the UK.

Subsequently, only two more flights were raised between 1947 and 1964. Very reluctantly, one Squadron HQ and another two flights were added in 1965. Planned growth at the scale of one Squadron HQ consisting of three flights for each corps started only in 1968/69. Although not upto full strength, the shortages were largely made up on the outbreak of the 1971 War.

J&K Operation, 1947-48

Air OP got the first opportunity to prove its mettle when tribals, covertly supported by the Pakistan Army, entered Jammu and Kashmir (J&K) in October 1947. Maharaja Hari Singh, who was still vacillating about his decision to join India or Pakistan, requested India for assistance to save his State but the Indian government made it very clear to him that it would intervene only if he signed the 'Instrument of Accession'.

On 27 October 1947, an Air OP aircraft from No. 1 Air OP Flight, piloted by Capt Sridhar Mansing (later Brig) flew Mr. Baldev Singh, Defence Minister, and Maj Gen K S Thimmaya from Jammu to Srinagar with the 'Instrument of Accession' which was signed by the Maharaja. A few hours later, the Indian Army landed its first unit in Srinagar. During the initial phase, Air OP pilots undertook a number of communication sorties to fly important Army and civilian VIPs between Jammu and Srinagar.

In January 1948, the flight was moved to Jammu. By this time, the tribals had also invaded the Jammu sector and were advancing towards Poonch, Naushera, Jhangar, Rajauri, Mendhar and Chhamb. Soon, it became an open war between the Indian and Pakistan regular Armies. Commencing with the familiarisation sorties flown for the Defence Minister S. Baldev Singh, Gen K M Cariappa and Lt Gen Kulwant Singh, the Air OP pilots undertook many reconnaissance missions and passed timely information about the enemy movements to our formations being deployed in this sector. The flight operated from the ALGs at Naushera, Jhangar, Rajauri, Mendhar and Chhamb which were within the Pakistan artillery range.

On 3 July 1948, Brig Mohammed Usman, Commander 50 (Independent) Para Brigade, who led from the front and fought a brilliant battle and stopped the enemy from capturing Naushera and Jhangar, was grievously wounded due to enemy shelling. Capt Sridhar Mansing who was tasked to evacuate Brig Usman, landed on the ALG which was under heavy shelling by the Pakistan artillery and evacuated him to Jammu. Unfortunately, Brig Usman who was awarded the Maha Vir Chakra (Posthumous) did not survive. Capt Sridhar Mansing was awarded the Vir Chakra for his daring and outstanding act of gallantry.

On 14 December 1948, Maj (later Brig) Shahane was especially tasked to locate and engage the enemy guns which were causing a lot of casualties on our troops. As he approached the area, he was greeted by anti-aircraft fire. Evading the hostile fire, he continued with his mission and located the guns which were engaged, all this in full view of our own troops who had all praise for him. He was awarded a well deserved Vir Chakra. His citation read:

> On 14 December 1948, Major Shivaji Wasudeo Shahane was detailed to go up in an aircraft to locate and engage enemy positions at Batot and Bagsar. As he flew over the enemy area, he was met by

heavy anti-aircraft fire, but without caring for his personal safety, he reconnoitered an enemy battery and engaged.

As observation was difficult, he came back and went up again for cross-observation. He was again greeted by anti-aircraft fire, but he neutralised the enemy battery. Later, he located two other gun positions.

Thus, by his dauntless courage, dogged determination and skill in airmanship, he carried out his mission successfully from an unarmed and unarmoured aircraft.

The Air OP pilots time and again ventured over the enemy forward localities and gun positions and directed own artillery fire on the enemy positions. In the battles of Kot and Jhangar near Naushera, our attacks had got stalled due to intense enemy artillery fire when a call was made for Air OP pilots who located and destroyed the enemy guns which resulted in the capture of two enemy posts.

With reinforcements reaching both sides and the Pakistan Army which was openly supporting the battle, the intensity of the battle increased, resulting in a high casualty rate. Air OP was constantly on demand to evacuate seriously wounded soldiers from the forward localities, landing on ALGs which were invariably under the enemy observed fire. Capt (later Lt Col) E K Harikrishan describes about two such typical missions thus:

> Two sorties stand out in my mind. The first, a Gorkha cook with his back blown off by a mortar bomb. He looked pale .... medical officer asked me to rush the patient off without even strapping him for the forty-five minute sortie ..... Half way through he suddenly woke up, sat half up, gazed at me with glassy eyes. My first thought, to protect my controls from a delirious grasp, my second, a numbing fear of someone dying besides me, helpless and alone by my side. My landing was not perfect but my "unbroken bungee cord" record still held. Before two months passed, I again flew this smiling Gorkha lad for his permanent

posting. My next casualty evacuation was that of a Subedar with a bullet in his chest touching his heart. He could not be flown at my normal mountain height. I flew him at deck level along the Tawi River all the way back to base, barring a climb over Akhnur Bridge. He survived my flight and his subsequent operation.

Thus, the young pilots of No 1 (Independent) Air OP Flight, who had very little flying experience and had not carried out any affiliation training with any Army formation, set the future standards by replicating the history of Air OP pilots in World War II in their own small way. In addition to two Vir Chakras, the flight also earned six 'Mention-in-Despatches.'

Hyderabad and Goa Operations

Soon after the ceasefire in J&K, trouble started brewing in Hyderabad where the Nizam—the ruler of Hyderabad state—stubbornly refused to accede to the Indian Union as he wanted to maintain his independent status. When all the diplomatic efforts failed to persuade the Nizam, who refused to see the writing on the wall, and also in view of the threats of a bloodbath by the Razakars, a militant organisation in the state, the Government of India decided to intervene in an operation called 'Operation Polo.' 1 Armoured Division under the command of Maj Gen (later Army Chief) JN Chaudhary moved into Hyderabad on 13 September 1948. No. 2 Air OP Flight allotted to the division for this operation was assigned the task to look for the dispositions of the Nizam's forces ahead of own advancing columns. Thus, the Air OP pilots were the eyes and ears of the advancing columns who were passed information on the radio and through message drops which speeded up the advance. The operation lasted only four days.

They also dropped leaflets over the headquarters of the Commander-in-Chief of the Hyderabad State Forces and the palaces of the Nizam and Crown Prince, asking them to surrender which they did. A time and place for the surrender ceremony was fixed, for which the GOC 1 Armoured Division got delayed. Due to the breakdown in communication, Air OP was tasked to drop a message to Maj Gen E L Edroos, Commander-in-Chief of the State Forces, informing him about the delayed arrival of Maj Gen JN Chaudhary for the surrender ceremony. In a hurry, Maj (later Brigadier) Govind Singh, Flight Commander, after take-off realised that he should have taken a written message from the Div HQ. However, he pressed on and on reaching overhead the surrender ceremony venue, he scribbled on a piece of paper "From Maj Govind Singh to Maj Gen EL Edroos. Stay where you are, General Officer Commanding 1 Armoured Division will meet you in due course." The message became the subject of much hilarity when Maj Gen EL Edroos returned that piece of paper to Maj Gen JN Chaudhary. Soon thereafter, No. 2 Air OP Flight pilots landed at Begumpet airfield and took over the Air Traffic Control, six Dakotas, two Chipmunks, two Beachcraft and some trainer aircraft of the Nizam's forces. The company of the Hyderabad State Forces deployed on the airfield was also disarmed.

Support given by Air OP pilots was highly commended by Maj Gen JN Chaudhary. It will be worth mentioning that Maj Gen JN Chaudhary, on taking over as Chief of the Army Staff (COAS) in 1962, was the first to take up the case for the formation of the Army Aviation Corps.

After the departure of the British from India, the Portuguese continued to rule Goa, Daman and Diu which were small enclaves on the western coast. When all efforts to resolve the issue peacefully failed, the Government of India decided to intervene and liberate these colonies by military intervention.

Maj Gen KP Candeth, GOC 17 Infantry Division, with 50 (Independent) Para Brigade placed under him, was assigned the task to capture these enclaves. No. 3 Air OP Flight located at Adampur (Punjab) was moved to Belgaum for this operation. Since the maps of the area were outdated, the first task given to the flight was the photo mission. Photographs of the bridges and the terrain around the two axes on which the advance was to take place were delivered to HQ 17 Infantry Division on 16 December and the Army columns moved in on 18 December. Air OP was tasked to pass information on the enemy movements and engage enemy armour and artillery emplacement in case of any resistance. With hardly any opposition, the flight pilots were bold and flew deep into the enemy territory well ahead of own advancing columns, passing them real-time information about the enemy, which speeded up the whole operation.

The fast pace of the operation caught the enemy unawares, forcing a quick withdrawal by the opposing force from Panjim and Marmagoa. On 21 December, the first Auster IX landed at Dabolin airfield and the whole flight concentrated there by 22 December. The entire operation lasted for 42 hours during which the flight flew 146 sorties in 88 hours of flying time.

The work of the flight was lauded by all the Ground Commanders and specially commended by Maj Gen KP Candeth who was appointed Lt Governor of Goa, Daman and Diu. Later Gen PN Thapar, the COAS and Lt Gen JN Chaudhary GOC-in-C Southern Command, complimented the flight when they flew down to Dabolin. The flight returned to Adampur (Punjab) on 5 January 1962.

1965 War

During the expansion phase of the Indian Army, even after the 1962 debacle, Air OP remained a low priority, probably due to the paucity of funds or lack of awareness about its potential during war. Till the end of 1964, the Air OP force level remained at one Sqn HQ and four flights. Then, very reluctantly, HQ 660 Air OP Sqn was raised on 1 January 1965, followed by, the raising of No. 5 Air OP Flight on 1 April 1965 at Nasik Road and No. 6 Air OP Flight on 1 July 1965 at Bagdogra.

But the raisings of No. 5 and No. 6 Air OP Flights were just notional as the resources of pilots and aircraft of No. 5 Flight were given to No. 1 Air OP Flight already deployed in the Rann of Kutch and No. 6 Air OP Flight got the aircraft only in July 1966.

Foreseeing the gathering storm, thanks to the initiative of Brig FSB Mehta, Brigadier Artillery HQ Western Command, who had done the necessary spadework in July/August, No. 7 Air OP Flight was raised at Patiala on 6 September and No. 8 at Jalandhar on 11 September by borrowing Pushpak aircraft from flying clubs in Punjab and reverting old Air OP pilots from the units. Lt Col DA Dhareshwar, commanding 660 Air OP Sqn, who had already done conversion training on Pushpaks, converted the pilots posted to these flights which were made operational within 15 days and launched into battle. Thus, the format of Air OP before the outbreak of open war was:

Eastern Sector – 659 Air OP Sqn with No. 4 Air OP Flight at Missamari.
Western Sector – 660 Air OP Sqn with No. 2 Air OP Flight at Jammu in support of 15 Corps.
No. 3 at Adampur in support of 11 Corps and No. 7 in support of 1 Corps.
Rann of Kutch – No. 1 Air OP Flight.

Thus, Air OP was launched into battle with big handicaps. The first was the unsuitability of Pushpak aircraft for Air OP. Sluggish at the controls and with limited space in the cockpit where the Army radio set was tied to the second seat, it had limited visibility to the front and no visibility to the rear. Even the Auster Mk V used in World War II was better than the Pushpak. Second, the pilots were straight sent to the battle without any affiliation training with the formations to be supported. Third, the pilots were imparted only inadequate and ad-hoc flying training on the Pushpaks.

This ad-hocism claimed its cost when Maj MS Bakshi, a brilliant officer under whom this author had served from 1955 to 1957, who was reverted back to Air OP from being GSO2 (Ops) of a division, died while approaching Rajasansi airfield.

But the major handicap was the inadequacy of Air OP resources. Only one aircraft could be allotted to each division instead of the whole flight of five aircraft later authorised after this war.

However, in spite of all the handicaps, Air OP pilots performed admirably in this brief war of three weeks.

Rann of Kutch

A number of events led to the massing of troops by India and Pakistan on the border in early 1965 and the subsequent open war in September 1965. The first flashpoint was the Little Rann of Kutch where a border post called Sardar Post, manned by the Border Security Force (BSF) was attacked by the Pakistan Army on 9 April 1965 when 31 Infantry Brigade Group, located at Bhuj, was ordered to take over the responsibility of the complete border of the Rann of Kutch. Subsequently, reinforced by more troops, a 'Kilo Sector' HQ was formed at Khavda under Lt Gen P O Dunn to conduct subsequent operations in the Rann of Kutch.

This author was then a Captain in No. 1 (Independent) Air OP Flight at Nasik Road. On 2 April morning, he was called by the Flt Cdr Maj (Later Maj Gen) SK Mathur and asked to fly to Bhuj as early as possible on that day and report to the Brigade Major, 31 Infantry Brigade Group there. He took off in the afternoon and after a night halt at Ahmedabad, landed at Bhuj on 3 April morning, to find that Major (later Major General) Rajinder Singh the Brigade Major himself was an Air OP pilot. Being fully aware about the range of Air OP and its capabilities, he had rightly taken the initiative to requisition the Air OP which could cover the vast area (about 5,100 sq km) in a single sortie that would take days for the ground troops to go over. Later, Capt GK Grover arrived with another aircraft on 11 April and the whole flight concentrated at Khavda by 21 April 1965.

The Pakistani artillery ground observers had a distinct advantage there as they had sand-dunes running parallel to the border on their side of the International Boundary (IB) and could observe our troops deployed in the open flat terrain.

There was very heavy demand on Air OP and two to three aircraft were always up from first to last light on reconnaissance and air shoot missions. Nothing could go unnoticed by the hawk-eyed, experienced Air OP pilots in this open sandy terrain.

On 23 April, Air OP reported movement of the enemy behind the sand-dunes towards Biar Bet which could not be observed by our ground OPs. On 29 April, the author, while flying his fourth and the last sortie of the day, reported the movement of Pakistani armour near Biar Bet and just a few minutes before last light, of six tanks crossing the IB and entering the open area of Biar Bet which he was asked to engage with 71 Medium Regiment. Observing that the first round had landed much short of the target, he ordered the guns to fire at the maximum range on the same bearing. To his dismay, the round was still short but the tanks stopped in their tracks. Seeing their reaction, he ordered firing of a few salvos which obviously were again short, and then returned to the ALG to land just at last light. The guns were ordered to redeploy forward at night.

Maj SK Mathur, Flight Commander, took off next morning to find that the tanks had withdrawn from Biar Bet to the positions behind the IB where he located more enemy concentrations which he engaged, destroying an ammunition dump. Maj S K Mathur who always led his pilots from the front, for his outstanding leadership was conferred the well deserved award of the Maha Vir Chakra. His citation read:

> Maj SK Mathur was in command of an air operation flight during the operations against Pakistani intruders in the Kutch area. He personally flew 45 operational sorties.

On 15 April 1965, Maj Mathur effectively engaged a convoy of intruders in the Kanjarkot area and destroyed three vehicles and damaged others. Again, on 30 April, he directed artillery fire towards the intruders in the Biar Bet area and compelled them to withdraw in haste. In this action, a field ammunition dump and three vehicles belonging to the intruders were damaged.

Throughout the operations, Maj SK Mathur displayed great courage, technical skill and leadership in the best traditions of the Army.

A very humble and likeable person, Maj Mathur said, "The award is for all of you and for the flight for its outstanding performance."

With ceasefire declared on 1 June 1965, operations in this sector came to an end.

J&K and Punjab

Capt CS Krishnan from No. 2 Air OP Flight was the first to be launched on 1 September in the Chhamb sector. He flew three sorties, totalling six and a half hours of flying, effectively engaging the enemy and passing accurate information about the enemy advancing columns. Later, he was continuously up throughout the war and was awarded the Vir Chakra. His citation read:

> On 1 September 1965, when Pakistani forces launched a massive attack in the Chhamb Sector, Captain Chitoor Subramanium Krishnan flew over the area in order to locate enemy artillery guns. Disregarding heavy enemy fire, he engaged the enemy artillery guns and ultimately silenced them.
>
> As officer commanding of the Air Observation Flight in Jammu and Kashmir, Captain Krishnan employed his flight in directing artillery fire against the infiltrators also and inflicted heavy casualties on them. He flew nearly 60 hours of which 45 hours were on operational sorties.
>
> Throughout, Captain Krishnan displayed courage and leadership in high order.

On 7 September, No. 2 Air OP Flight was also given the responsibility of the Sialkot sector and plains sector of Jammu. On 8 September, Maj MM Bapna and Capt Virendra Singh, in two separate Auster Mk IX, were sent on special photo missions over the enemy held area. They completed the missions but suffered substantial damage to their aircraft due to enemy fire. However, with skill and presence of mind, they landed back safely.

On 13 September, Capt MM Berry, while on a reconnaissance and air shoot sortie, was hit by enemy fire and had to land his Pushpak in the 'no man's land' in the Sialkot sector. He was rescued but severely injured and was evacuated to the Military Hospital, Udhampur, in time and survived.

On 13 September, Capt HS Choudhary was again hit by enemy ground fire while directing artillery fire in the Sialkot sector and had to crash-land. On this day, three other aircraft of this flight got damaged due to heavy enemy shelling on the ALG. But the ground crew rose to the occasion and repaired the damage. Air OP pilots were in great demand and, at times, all the aircraft were up to cover this extended sector. The last day of the war was the day of 'air shoots' when a large number of air shoots were undertaken to engage enemy tanks and vehicle columns.

In addition to Capt CS Krishnan, Capt Verinder Singh, Capt MM Berry and Capt JS Nakra were awarded Sena Medals (Gallantry) and Maj MM Bapna was awarded both Vishist Seva Medal and Sena Medal (Gallantry).

Pilots of No. 3 Air OP Flight which covered the whole frontage of 11 Corps in Punjab, with only one aircraft supporting a division equally distinguished themselves. Air OP pilots had to contend with hostile F 86 Sabres which were particularly active in support of their troops in this sector.

Although it was amply proved in the earlier incidents that unarmed Air OP aircraft could survive even against the most hostile ground and air environments, those sceptics, both in and out of uniform, who still feel that an Air OP is a sitting duck, would do well to read the citation for the Vir Chakra awarded to Maj (later Brig) SS Ratra, also a paratrooper, commanding No. 3 Air OP Flight:

> Maj SS Ratra was an AOP officer in the Lahore Sector from 15 to 23 September 1965. To gain vital information and direct accurate fire on Pakistani tanks concentration, bridges and crossing sites over the Ichhogil Canal, he had to undertake a number of flights over enemy picquets. With great courage, he carried out these extremely dangerous missions in an unarmed aircraft. Based on his information, our aircraft destroyed an important enemy bridge and, thus, cut off the enemy's armour exit route. He was responsible for the destruction of five enemy tanks and several vehicles. On numerous occasions, while flying low, he was fired at by enemy medium machine guns. On 17 September, he was attacked by an enemy F 86 fighter and on 21 September, he was encircled by three F 86 fighters which fired thrice but on all those occasions he kept his composure and came back safely.

Another Vir Chakra award in this sector went to Capt (later Maj Gen) SC Sabharwal, a cool, determined, dedicated and very professional Army aviator with whom this author had the opportunity to serve during their Air OP tenures.

1971 War

The Indo-Pak War of 1965 brought out the glaring shortages in the force levels of Air OP. Based on the lessons learnt and after long debates with the Air Force, the planned growth of Air OP – one Sqn HQ consisting of three Air OP flights for a Corps HQ started in 1968/69. HQ 661 OP Squadron was raised in January 1967, followed by the raising of No. 10 and 11 Air OP Flights in 1969. This was followed by the raising of No.12 (Independent) Air OP Flight in 1970 and No. 14 and 15 Flights in 1971.

This expansion phase had its own pains. Although the problem of pilots was overcome by conducting an ad-hoc course at the

Flying Club, Patiala, under the aegis of 660 Air OP Squadron, the equipment problems were quite serious.

Production of Auster Mk IX, an excellent fixed-wing aircraft for Air OP, was stopped by the UK in 1960. Thus, the Auster serviceability in the existing Air OP Flights was very low. A Qualitative Requirement for its replacement had earlier been given to Hindustan Aeronautics Limited (HAL), Bangalore. The 'Krishak' produced by HAL was inducted in Air OP in 1966 but proved a very poor substitute for the Auster Mk IX. At the same time, the Army also was pressing for the induction of helicopters which, after the initial reservations by the Air Force, was agreed to in 1968. Thus, on the outbreak of hostilities, although not full strength, Air OP resources, a mixed bag of Auster IX, Krishaks and Chetak helicopters, allotted to the formations were: each Corps HQ allotted a Sqn HQ consisting of two to three flights, and some divisions deployed independently, allotted one Air OP Flight. Since Austers and Krishak flights did not have good serviceability, the Pushpaks of the flying clubs again came in handy and were allotted as replacement and war reserves to the fixed-wing flights.

Since the aim of this book is only to give an actual account of the Air OP, let us see how they replicated the battle accounts of their predecessors in the 1971 War.

Shakargarh Sector

660 Air OP Sqn commanded by Lt Col (later Brig) NR Pawar, an officer of old vintage with No. 1 Air OP Flight commanded by Maj (later Lt Col) SY Rege (a known chronic bachelor who found a life partner after retirement) and No. 9 Air OP Flight, commanded by Maj Vikas Kapoor, supported various formations in this sector where the Pakistan Air Force was very active in the first few days. On 5/6 December, there was a lot of ambiguity about the correct positions of own troops in the 39 Inf Div sector which Air OP was asked to check. In view of the importance of this task, Lt Col NR Pawar himself undertook this mission on 6

December and cleared the confusion by reporting their correct positions. The same day, Maj SY Rage flew a sortie in the 54 Inf Div sector, engaged a column of armour, resulting in the destruction of two tanks.

On 9 December, Capt DPS Nurpuri took Maj Chaudhary of 4 Horse for reconnaissance and located sizeable enemy armour across the border. As a result of this vital information, own ground commander modified his offensive plan, facilitating a successful manoeuvre by our armoured columns.

Another high point of Air OP in this sector was on 14 December when Capt PK Gaur and Capt GS Punia took off to register targets for an attack planned in the afternoon. By then, Air OP had become quite a nuisance for the enemy who started mounting special Sabre missions to take care of them. As they were registering the targets, four Sabres swooped on them. They managed to dodge the Sabres by taking evasive action but continued with their mission. However, a little later, the Sabres appeared again, fired a rocket which again missed the Krishak but the other Sabre fired a long burst of cannon which hit the fuel tank, setting the Krishak on fire and killing Capt PK Gaur instantly. Showing remarkable presence of mind, Capt GS Punia, with the controls partially effective, managed to crash-land the burning aircraft behind own forward defended localities. Capt GS Punia suffered serious third degree burns but survived. For his supreme sacrifice, Capt PK Gaur was awarded the Maha Vir Chakra (Posthumous) and Capt GS Punia, the Vir Chakra.

In this sector, enemy artillery would also fire airburst shells to scare the Air OP pilots. At the start of the war, Capt Madan Pal's aircraft got a hit by an airburst splinter but no harm came to him. On 14 December, he was hit again by an airburst splinter, damaging two pullies of the elevator cable but again managed to land back safely. The third time he was not as lucky when he had taken along Capt Sapru Mathew, who was still not fully qualified, to undertake operational sorties. After completing their first task which was registration of targets with the guns of 41 (Independent) Artillery Brigade for an attack on Shakargarh, they were tasked to look for the enemy in another area where an enemy MMG fired on them. Capt Madan Pal was hit on the chest and immediately slumped in his seat. Capt Sapru Mathew took over the controls and in spite of his limited experience, flew back, force-landing in the 'no man's land' in front of 22 Punjab in full view of the enemy who again brought down artillery and small arms fire on them. Capt Mathew managed to take out the body of Capt Madan Pal and came to a safe place. For his supreme sacrifice, Capt Madan Pal was awarded the Vir Chakra (Posthumous) and Captain Sapru Mathew the Sena Medal (Gallantry).

Rajasthan Sector

No. 5 (Independent) and No. 12 (Independent) Air OP Flight were placed in direct support of 11 Inf Div and 12 Inf Div under HQ Southern Command located at Jodhpur. 11 Inf Div was deployed in the Barmer sub-sector and 12 Inf Div in the Jaisalmer sub-sector. Both these flights had moved from Nasik Road and had neither done any affiliation training with these formations nor operated in the desert terrain earlier. Both were equipped with the Krishak aircraft, which due to the sand ingress through the air-intake facing downward gave a lot of problems, resulting in forced landings. In anticipation, these flights were given three Pushpaks each as war reserves.

Both these formations were given the task to move deep into Pakistan on the outbreak of war.

11 Infantry Division

On the very first day of the war, Capt P Madhavan was tasked to locate the enemy forward positions. He was engaged by enemy ground fire and hit, shattering his aircraft canopy and damaging his radio set, cutting off his communications. As he turned, he also found that his control column which had also been hit, was partially effective but he managed to land safely on an unprepared ALG on own side of the IB. After repairs the next day, he was again up on Air OP tasks. The same day, Capt Yusufji, while on a similar mission, was hit by ground fire, suffering partial engine failure but he managed to land back safely. Next day, while on an air shoot mission, he was engaged by anti-aircraft guns and later chased by Sabres which he managed to evade and landed back safely on the ALG.

By 8 December, this formation which had made a rapid advance, was planning to attack Naya Chor deep inside the enemy territory. Capt Madhavan and Capt Rajan were given the task to search for the enemy and drop a message to own troops about the enemy locations. After he had completed the task, the report came that the aircraft was missing and the search mission launched in the limited daylight hours could not locate them. As known later, Capt Madhavan had to force-land after experiencing engine failure but managed to save himself and Capt Rajan. Thereafter, they trekked back about 20 km over the sand-dunes, at times taking a camel ride, and reached the base next morning. Next day, they resumed Air OP missions although they were a bit uncomfortable in the aircraft seat for a few days as an after-effect of the camel ride.

The pilots of this flight, under the leadership of Maj KB Deswal, Commanding Officer, displayed ingenuity, a high level of skills and initiative throughout the operations which is evident from another incident.

On 16 December, Capt (later Brig) SN Gupta and Capt P Madhavan, when on a special mission, experienced engine failure. They turned back but soon had to force-land in the occupied territory. As they were removing their maps and the survival gear, they saw a group of locals from a Pakistani village approaching them. Keeping cool and with presence of mind, Capt Gupta told them to halt at a distance and asked the oldest man in the group to come forward. Capt Gupta then told the old man that the Indian Army had captured Naya Chor and the Indian troops were already beyond their village and there was a powerful bomb in the aircraft which could blow up their whole village. To convince the old man, Capt Gupta switched on the Radio Set ANPRC 25 and made the old man hear the mushing sound. The old man was told that he must look after the aircraft for the sake of the safety of his village. From the villagers, Capt Gupta then got a camel and a guide and they found their way back to a BSF post. After the ceasefire, when Capt Gupta came back to retrieve the aircraft, he found the old man sleeping under the shade of the aircraft, personally guarding it from any harm. The aircraft was dismantled and recovered partly by camels and partly by a MI-4 helicopter.

In spite of many engine failures, the maximum in a flight in the 1971 War, No. 5 Air OP Flight logged over 230 flying hours in 227 sorties. Capt P Madhavan and Capt IA Yusufji were awarded well-deserved Vir Chakras.

12 Infantry Division

12 Infantry Division was deployed in the offensive role and tasked to capture Rahimyar Khan with a view to disrupt the rail-road link from Lahore to Karachi. The division was concentrated in the area of Kishangarh short of the IB to advance at first light on 5 December 1971. No. 12 Air OP Flight, commanded by this author, then a Major, was in direct support of this division. On the night of 4/5 December, Pakistan preempted this plan when a Brigade Group led by a Regiment of Armour crossed the IB 70 km south of Kishangarh and headed towards Longewala 16 km inside the IB which was held only by an infantry company. 12 Infantry Division and HQ Southern Command were completely taken by surprise as, in the initial appreciation based on intelligence reports, such a threat was ruled out. Initial reports by a strong patrol sent to the IB were not believed but when the patrol leader said he could see the tank column moving in front, the real action started. The advance to Rahimyar Khan was put on hold and the GOC Maj Gen RF Khambatta told the nearest Brigade Commander at Sadhewala 20 km north of Longewala, to send reinforcements. This was around 0345 hours. He also contacted Wg Cdr (later Air Marshal) MS

Map 8: 12 Infantry Division

Bawa, Station Commander Air Force Station Jaisalmer, who would send the Hunters at first light as the Hunters had no night attack capability.

No. 12 Air OP Flight was tasked to act as airborne Forward Air Controller (FAC) for the Hunters. The flight was deployed on the ALG at Ranau, 15 minutes flying time to Longewala.

Initially, everything seemed to work to the advantage of the enemy who, however, had not realised the problem of armour and vehicles moving 16 km cross-country through the soft sand-dunes. A few hours late in their time plan, they could contact Longewala only at first light. Capt (later Col) PS Sangha was launched at first light on 5 December to act as airborne FAC to the Hunters. The first Hunter mission, consisting of two aircraft, reached overhead Longewala just minutes before Sangha. After hitting two tanks,

Krishak Aircraft at Longewala 1971

they got doubtful and thought they might have hit our own tanks. Capt Sangha, monitoring their R/T communication, told them to wait. In two minutes, he was overhead Longewala, flew over the battlefield, identified that all these were enemy tanks and told the Hunter mission to get on with them.

In this sortie of two hours, Sangha directed two Hunter missions, getting six tanks destroyed. Between the Hunter mission, he would go around the flanks of the enemy to pick up more targets as by then he had identified the enemy axis of advance. He was fired upon by the enemy every time he flew beyond Longewala, at times, his observer, Raj Singh telling him, "Sir, we are being fired from there." Being very experienced and a Flying Instructor, his flying skills were tested to their limits.

He also would fly back to the reinforcing column which was still quite short of Longewala and inform them about the latest situation by dropping messages as he had no radio communication with this column. In his third message, he warned our armoured

The Victors of Longewala, Capt PPS Sangha, Maj Atma Singh and Flt Lt CP Naidu (Flight Adjutant) at forward ALG in Jaisalmer Sector

corps officer leading a troop of four AMX 13 tanks about two enemy tanks in hull down positions awaiting to 'receive' him.

Running low on fuel, he landed at the ALG and asked the Flight Commander who was ready for the next sortie, to allow him to go for the next sortie as "being in the full picture, will do the job better and also there was no time to brief, since the next Hunter mission was expected any time." Within five minutes of his landing, he again took off and was soon overhead Longewala to await the next Hunter mission. He also observed that in spite of his timely warning to the armour officer, two of our AMX 13 tanks had been knocked out.

Soon after Sangha had taken off, the GOC got Maj Atma Singh on the phone and asked him to look for the enemy tanks which according to ground reports had bypassed Longewala and were heading towards Ramgarh or Ranau. He immediately took off and searched the complete area of about 1,700 sq km, forming a triangle between Longewala, Ramgarh and Tanot. Making doubly sure that no tanks had bypassed Longewala and there was no threat to Ramgarh or Ranau, he flew straight to Tanot, and passed this information by dropping a message on Div HQ. He landed back at the ALG at 1235 hours.

In the meantime, Sangha, during second sortie, fully in command of the situation at Longewala, directed two more Hunter missions and got another eight tanks and scores of vehicles destroyed. He also kept our reinforcement which had arrived on the post around 1130 hours, updated with the latest situation.

The enemy was now in tatters and had no place to hide as each of his movements was picked up by Sangha and reported to the Hunters. The enemy had frozen all movement and the only thing seen moving by Sangha were the plumes of smoke emitted by burning tanks, looking like wind socks. He was told by the last Hunter mission that the next mission would be overhead Longewala at 1300 hours. Thus, an unarmed aircraft flown by an unarmed pilot had created havoc at Longewala. Sangha landed back at Ranau at 1215 and waited for Maj Atma Singh who soon after, landed back after his first sortie.

Sangha gave a detailed briefing to Maj Atma Singh and also told him that there were still many more tanks intact which he had seen when they were on the move. Maj Atma Singh took off for another crucial sortie at 1245 hours but soon after that, the GOC Maj Gen RF Khambatta arrived at the ALG. Despite the detailed information passed about Sangha's two sorties and the dropping of a message on Div HQ by Maj Atma Singh, he was still getting contradictory reports from the ground troops, creating a lot of confusion. So he decided to drive all the way from Tanot to Ranau, a distance of 21 km. He told Sangha that he was getting all sorts of reports from the ground troops but he knew that the best and most trusted information came from Air OP and had decided to listen to him personally. Sangha gave him a detailed debrief of his two sorties. The GOC spent a good 30 minutes at the ALG and left quite relieved and at peace.

Maj Atma Singh, after reaching overhead Longewala at 1300 hours, could only see the plumes of smoke from the burning tanks as the enemy, now under camouflage, had frozen all its movements. There was an occasional explosion from the shells of burning tanks which would shake the fabric of his Krishak. As the Hunter mission was late, he started looking for the tanks which Sangha had said were still intact. Increasing his height to remain outside the enemy ground fire range, he flew about 8 km into the enemy held area when suddenly a tank disclosed its position by firing at him. On close scrutiny, he located seven tanks under camouflauge in that area. He turned back and soon after contacted the Hunter mission which was directed on these tanks. A few minutes later, he experienced loss of engine power, with the aircraft gradually losing height. He noticed that the oil temperature had gone beyond the red line and the oil pressure below the limit, indicating oil leakage in the oil system. He had to make a quick decision to either force-land behind the Longewala post in the sand-dunes or alternatively on the helipad which by then was in the 'no-man's land'. He opted for the latter and after giving a call to the Hunters, landed safely on the helipad. The Hunters had assured him that they would take care of any enemy coming towards him. After switching off, he went to an abandoned tank only 15 yards away which was the enemy Squadron Commander's tank and removed its pennants, a map fully marked with the attack plans and two steel helmets left behind by the tank crew. Along with his observer Raj Singh, he walked back 700 metres to the Longewala post. After destroying all the seven tanks and seeing him safely back, the Hunters returned to the base.

At 1500 hours, when Major Atma Singh was still waiting for a jeep to take him back, he saw another Hunter mission overhead but there was nobody to guide it. He rushed back to his Krishak on the helipad, opened its R/T and told the Hunters to take on whatever they saw beyond him and beyond the Longewala post. He also got the forward most jeep of the enemy destroyed as he had earlier seen it during his final approach to the helipad.

He returned to the post where a jeep had arrived for him and on the way back, first briefed the Brigade Commander at Sadhewala and then briefed the GOC in detail at Tanot. He also gave him the enemy map taken from the enemy tank.

Maj Atma Singh then took his ground party from the ALG and went back to Longewala and retrieved his aircraft from the helipad after last light. By then, our artillery and reinforcements had arrived from Kishangarh and were registering enemy targets for the night. The Krishak was brought 5 km down the road by man-handling. Next day, technicians replaced the oil and the oil pipe and declared the aircraft fit for flying. Maj Atma Singh took off from the road in the afternoon and landed back at the ALG.

Thus, Air OP in this battle replicated the history of World War II by acting as airborne FAC to the fighters in close support.

5 December was the most fateful and the longest day for the Air OP pilots of No. 12 Air OP Flight which together with Hunter pilots, claimed the destruction of 21 T-59 tanks and scores of vehicle, resulting in complete decimation of the Pakistani armoured thrust. As learnt later, Pakistan had ordered withdrawal of troops at around 1600 hours that day.

The same evening, the GOC sent the order that from the next day, Air OP would keep one aircraft up from first to last light.

As the ground OP could not observe beyond the next sand-dune, from 6 December, Air OP pilots, besides directing artillery fire and Hunter missions on the retreating enemy, would pass very accurate and real-time information on the artillery radio set, first to the ground troops and then to Div HQ after each sortie.

After the ceasefire, two air OP pilots from the flight took a count of the losses of the enemy who left behind 27 tanks, 122 vehicles, 3 anti-aircraft guns and 3x25 pounder guns which were later retrieved.

The performance of the flight was personally commended by Lt Gen GG Bewoor, GOC-in-C Southern Command, during his visit to Longewala. Maj Atma Singh and Capt PS Sangha were awarded Vir Chakras. Capt Dinesh Mathur, Cpl Hariharan, Cpl Baladhan Dayuthan and Cpl Balan Pillai were 'Mentioned in Despatches.'

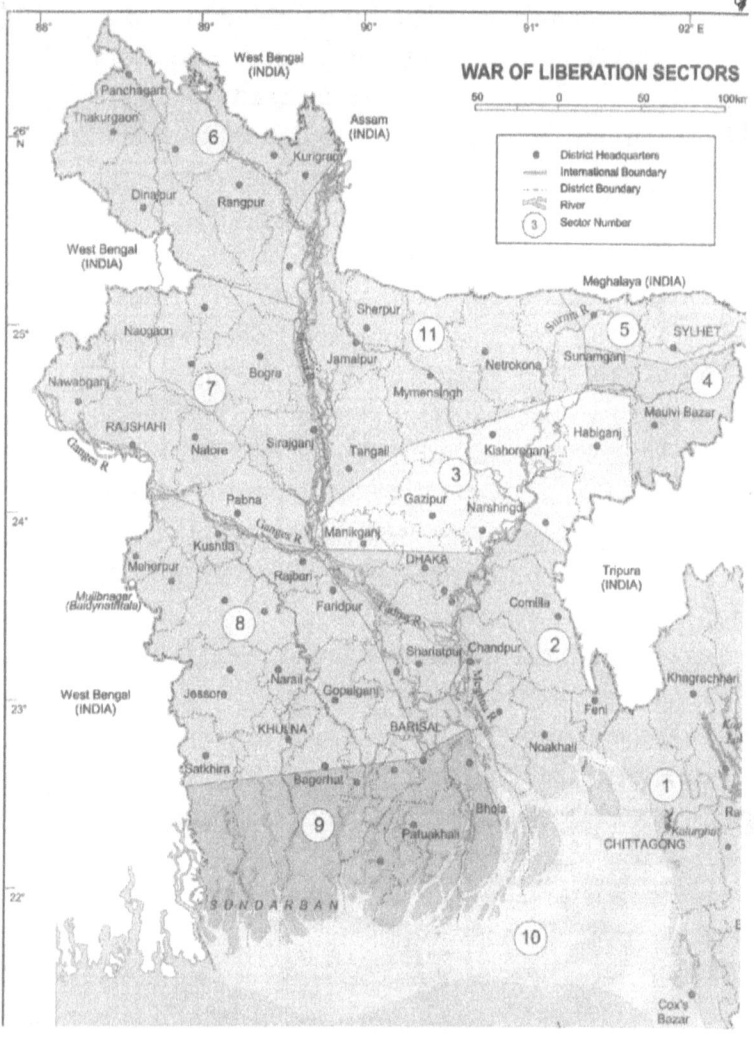

Map 9: Eastern Theatre

Eastern Theatre

659 Air OP Squadron, commanded by Lt Col (later Brig) SL Tugnait, based at Bagdogra, provided Air OP support in the Eastern Theatre. It had five Air OP Flights under it: No. 4 and No. 6 Flights equipped with Krishak aircraft and No. 10, No. 11 and No. 15 Flights equipped with Chetak helicopters. While the Squadron HQ remained at Bagdogra, the flights were deployed in support of the formations of 33 Corps, 2 Corps and 4 Corps.

When the balloon went up, Air OP was in great demand as the ground OPs had very limited observation in the flat obstacle ridden riverine terrain where Air OP could be exploited to its full potential. Air OP pilots were in the forefront and often given reconnaissance missions and other tasks which required them to fly deep into the enemy territory.

On 2 December, Maj EDE Menezes, commanding No. 6 Air OP Flight, while on a reconnaissance sortie, spotted enemy movement which seemed to be forming up for an attack against the 4 Guards defended positions. He engaged the enemy with own artillery, inflicting heavy causalties, frustrating the enemy plans for attack.

On 9 December, Maj Gen BF Gonsalves, GOC 57 Infantry Division, himself an Air OP officer, personally briefed Capt (later Lt Gen) G S Sihota to recce a suitable landing ground deep inside the enemy territory for a highly secretive heliborne landing operation across the Meghna river. The area of landing was to be such where the enemy could not react before our troops had gained a strong foothold. Although fired upon and hit by the enemy, he successfully executed the task by selecting a most suitable landing ground. In the afternoon, Capt G S Sihota led ten MI 4 helicopters with elements of 4 Guard and a troop of 151 Light Battery from 82 Light Regiment to the landing area. The operation was successfully executed.

In this operation, Capt G S Sihota was awarded the Vir Chakra. His citation, given below, speaks for itself:

On 9 December 1971, Capt Gurbaksh Singh Sihota was ordered to carry a reconnaissance party for the selection of a suitable landing site for a helicopter-borne operation in the Eastern Sector. Skillfully piloting his helicopter, he penetrated deep behind enemy occupied territory. During this reconnaissance, the helicopter was fired upon and hit by enemy small arms fire. Capt Sihota, however, brought the damaged aircraft safely back to a forward helipad. Although his helicopter was damaged, he undertook a mission to evacuate two serious casualties. Later, the same afternoon and in the same damaged helicopter, he led the first wave of the helicopter-borne operations and directed other helicopters to a safe landing.

Throughout Capt Sihota displayed courage, initiation and professional skill of a high order.

Subsequently, Air OP assisted in another heli-landing involving 41 helicopters which carried 65 Mountain Regiment and another Battery of 82 Light Regiment. This resulted in early capture of Narsingdi which opened the axis for speedy advance to Dacca.

Brig RC Butalia, Commander 6 Mountain Artillery Brigade, himself an Air OP pilot who had done his flying training in the UK and later commanded 660 Air OP Squadron at Patiala where he conducted an ad-hoc flying course for 35 officers in 1967 during the expansion phase of Air OP, made the optimum utilisation of Air OP resources under him. As a matter of SOP, he would send them first for reconnaissance and air photo missions of the objectives before the attack and also himself monitor Air OP broadcasts during the attack. During one such monitoring, Brig RC Butalia placed the complete artillery brigade at the disposal of Maj (later Brig) KK Mittal who had spotted an enemy column of tanks, guns and infantry vehicle which was engaged, inflicting heavy casualties.

Later, Maj KK Mittal effectively engaged with guns of 98 Mountain Regiment a bridge strongly held by the enemy who was holding up the advance of 71 Mountain Brigade, thus, speeding up the advance of this formation.

In another sector, Maj HS Wadalia and Capt RK Gogna, both later Brigadiers, took battalion and company commanders for aerial reconnaissance and also provided them with the photographs of the objectives before the attack which was successful. On 14 December, they located an enemy column with tanks from the direction of Dinajpur which was immediately engaged by them with medium artillery. They also directed the enemy fighters on the enemy which was moving towards a bridgehead held by our troops.

Bogra, held by Pakistan 16 Infantry Division and 205 Infantry Brigade, was another strong point on the way to Dacca. By last light 13 December, 340 Mountain Brigade had contacted its defences. An attack was planned next day at 0900 hours. Maj J M Agnihotri allotted 64 & 97 Mountain Regiments, and 44 Medium Battery first degraded the defences, registered the targets for the fire plan and then observed the artillery bombardment during the attack and also directed our fighter mission on the enemy defences. While flying south of Bajra, he saw an enemy column escaping from Bogra defences. This information was passed in time to 69 Armoured Regiment which had established a roadblock south of Bogra. He also engaged this column with 44 Medium Battery.

No. 10 Air OP Flight, equipped with Chetak helicopters, commanded by Maj (later Maj Gen) GS Hundal was placed in support of HQ 2 Corps. Very experienced in both on fixed-wing aircraft and helicopters and a qualified Flying Instructor, he brought up all his pilots to the fully operational standard in a short time. With four serviceable helicopters, he fully met the demand of Corps HQ and its formations, 4 Mountain Division and 9 Infantry Division, in the Jessore sector.

On 7 December, along with Capt (later Lt Gen) SJS Saighal, Maj Hundal undertook a reconnaissance mission to locate the enemy positions in Jessore which was believed to be held by the Pakistan Army in strength but found the enemy withdrawing from there. He immediately engaged the withdrawing column, inflicting heavy casualties. He also passed the information that Jessore airfield was clear of any enemy and landed there. A few minutes later, Maj Gen Dalbir Singh, GOC 9 Infantry Division, along with his Brigade Commander landed there in another helicopter.

On 9 December, along with Capt PR Misal, he undertook a reconnaissance mission in 4 Mountain Division Sector with Lt Gen TN Raina GOC 2 Corps, Col SF Rodrigues, his Col (GS) (later both became Army Chiefs) and Col SP Joshi from Army HQ on board over Kushtia which was strongly held by the enemy. On locating the enemy, without losing any time, he gave the fire orders and engaged the enemy, with all of them on board. The information about the enemy deployment was also passed to the Brigade Commander advancing to Kushtia.

In another mission in 9 Mountain Div Sector, Capt SJS Saighal and Capt MS Dullat, while on a reconnaissance mission with Brig KK Tiwari, Commander 32 Infantry Brigade and his supporting artillery Commanding Officer Lt Col MD Anand, were fired upon by an MMG. A bullet pierced the heart of Lt Col MD Anand who died instantaneously. MMG bullets also damaged the rotor blades but Capt Saighal flew back and landed the helicopter safely.

Maj GS Hundal was awarded 'Mentioned in Despatches' and Capt MS Dullat, the Sena Medal (Gallantry).

Air OP aircraft returning to the ALG after completion of the mission

Frozen Highest Battlefield

The Saichen Glacier, at heights ranging from 18,500 ft to 24,300 ft where the Indian Army has been fighting a 'peace-time hot war' could be called the highest battlefield in the world. With temperatures going down as low as minus 35 degrees Celsius and blizzards lashing up to 150 knots per hour, it has very low levels of oxygen for human survival. The region remains covered with snow throughout the year, creating 'white out' conditions for the pilots.

Till 1984, it was lightly held, but then it was decided to reinforce and occupy more posts on the ridge line.

663 Air OP Squadron, commanded by Col GS Hundal with No. 3 and 19 Air OP Flights under him based at Srinagar, was allotted to support the operation called 'Meghdoot'. Later, to meet the heavy demand to support this operation, No. 30 Air OP Flight from Pathankot and helicopters and pilots from other flights in the plains were given to him, raising his fleet from 10 to 27 helicopters.

Till then, none of the pilots had been trained to operate in the glacier area where pilots and machines had to fly to their limits and land on carpet size snow helipads in 'white out' conditions. Each pilot had to be cleared on each helipad, most of which had no margin for error and also could come under the enemy fire. Therefore, the first task of the Squadron Commander was the operational training and clearing each pilot to fly and land in the glacier area. Himself, an experienced pilot on both fixed-wing aircraft and helicopters and also a Flying Instructor, he completed this task and was ready to undertake the daunting task when Operation Meghdoot was launched in 1984.

Since 1984, hundreds of missions have been undertaken by Air OP pilots in unarmed Cheetahs in Operation Meghdoot which is continuing. But as this story of 'Unarmed Air Observers' is to end on 1 November 1986 when the Army Aviation Corps was formed, a few daring missions undertaken till then by the Air OP pilots in unarmed Cheetahs are worth mentioning to show their grit, determination, fearlessness and instinct to survive in the most inhospitable terrain in the world.

On 28 June 1984, Col GS Hundal, along with Maj PJ Kumar, undertook a reconnaissance mission in support of our troops

Helicopter landing on a Snow helipad

tasked to occupy a ridge line. As our troops were still short of the ridge line, they saw the enemy approaching the same area which they engaged, inflicting heavy casualties on the enemy who withdrew, enabling our troops to occupy and consolidate on the ridge line.

On 30 August 1985, Capt VS Guleria who was already a veteran in Siachen flying, along with Maj SK Gadhok, recently inducted, were tasked to engage the enemy targets across the Saltoro ridge at 18,000 ft. After reaching the ridge, he climbed higher to locate the enemy positions. Gadhok was handling the radio set and passing fire orders to the guns which soon reported ready. As Guleria turned, he was riddled with bullets from a hostile MMG. Gadhok was hit and with his hands still clinching the handset, slumped forward. The Cheetah's bubble canopy was splattered with blood. Guleria freed the handset from Gadhok and passed the order for fire to the guns.

Guleria then headed back and had an agonising 12 minutes of flight to the base camp knowing that not everything was right with the helicopter. On seeing the helipad, he decided to put down quickly the helicopter when the inevitable happened. He was still at a height when the helicopter's main gear box seized. He flared the helicopter which settled just short of the helipad to discover that he had flown with an almost dry gear box, as one of the bullets had ripped off the oil cooler seal, draining out all the oil.

Gadhok had died while passing fire orders to the guns and made the first supreme sacrifice by an air observer in Operation Meghdoot.

Both Maj SK Gadhok and Capt VS Guleria were awarded the Shaurya Chakra.

After the Commander was briefed by Guleria, it was decided that the post from where fire had come had to be destroyed.

Guleria was airborne early next morning now as a co-pilot to Maj Michael Anthony Pereira. They were again fired from the same post but they had planned for it and evaded the fire. They engaged the enemy with guns deployed in the base camp and destroyed the post.

Again, on 5 April 1986, while on a supply mission, Maj VS Guleria noticed no enemy movement on the 'saddle post' next to own Bilafond La post. To further make sure, he flew closer and was surprised to see not a single enemy on the post. He passed this vital information to the base camp. The matter was referred to Division, Corps and Command Headquarters. It was decided to occupy the post, which was also assisted by the Air OP pilots by taking troops in ones and twos in the Cheetahs, the only helicopter which could land there.

Another daunting and most challenging mission was undertaken by Col GS Hundal, Lt Col Sita Ram and Maj MS Lele in three helicopters. The Army Comamnder, Lt Gen BC Nanda with his Chief of Staff Lt Gen PN Vadhera and Corps Commander Lt Gen Sami Khan, after a meeting at base camp, decided that the mission, irrespective of the risk involved, about which Col Hundal had apprised them, had to be undertaken. The mission was to replenish ammunition, kerosene and rations to our Amar and Sonam Posts which were running low on these items. These posts were dominated by a nearby Pakistani post which interfered with our air supply. Two dead bodies were also to be evacuated from these posts. The mission was so important that Lt Gen PN Vadhera decided to go up in one of the helicopters to see for himself the execution of this mission. Col Hundal did meticulous planning and gave a detailed briefing to each pilot, accounting for all the contingencies that might arise during its execution. Leading from the front, he tasked Maj Lele to bring down artillery fire on the nearby enemy post and he himself and Lt Col Sita Ram, in two separate helicopters decided to go to Amar and Sonam Posts to deliver the essential supplies and evacuate the dead bodies. The mission was successfully completed. The Army Commander and Corps Commander had stayed back at the base camp till all the helicopters returned.

Author with Brig HS Sihota, Brig FSB Mehta,
Brig Govind Singh and Maj Gen MS Chahal
(1978 at Nasik Road)

Col GS Hundal, for his outstanding overall performance as a Squadron Commander, was awarded the Ati Vishist Seva Medal twice and Chief of Army Staff Commendation Badge. Lt Col Sita Ram and Maj VS Guleria were awarded the Shaurya Chakras and Maj Lele was 'Mentioned in Despatches'.

6

Postscript

In war, every new idea has its opponents and sceptics but there are men with vision who push through the new ideas when new tools and instruments of war are available, and they not only become the pioneers of the new concepts of warfare but also change the course of history. One such small idea came to de Villette, a Frenchman, when he was taken up in a balloon on 17 October 1783. His idea that a balloon could be used as an aerial platform to observe enemy positions and movements caught the imagination of some French military minds at a time when the French Army was confronting its hostile neighbours, Austria and Prussia. Thereafter, the balloons were used by the French Army and subsequently in the American Civil War quite effectively but the British Army resisted the employment of balloons as an aerial platform for observation and direction of fire for a long time.

Similarly, when the first aeroplanes by the Wright Brothers in the USA and their counterpart in the UK were offered for Army trials for reconnaissance, the cavalry which till then was used for the same purpose, opposed it, as the noise of these machines would frighten their horses and many others within the Army thought that it was a waste to spend time and money on the new flying machines which were very fragile and looked like mosquitoes and flies. But the air observers, both from the balloons and aeroplanes, not only proved their worth but were very effective and, at times, the only means to pick up information about enemy deployment and engage targets behind the enemy defended localities in World War I. Subsequently, this experience brought about revolutionary changes in the concept of warfare during the next two decades.

Every battle and war also brings out some lessons. Two major lessons were learnt from the experience of unarmed air observers from the balloons and aeroplanes. First, air observation is an art which only a ground soldier trained to identify and distinguish war-like objects, could learn and undertake. Second, the observers with artillery experience performed their tasks of observation and direction of fire more effectively. This was advocated even by the Royal Engineer officers who commanded the Balloon Battalions, later taken over by the Royal Flying Corps (RFC) before World War I.

After the war, all the flying machines came under the control of the Royal Air Force in the UK and the Air Corps in the USA, depriving their Armies of this element of air observation.

However, when the military strategists started evaluating the impact of air power on the ground battle, they realised that integration of certain elements of air power with the Army was not only essential but a great battle winning factor because in the ground battle environments, it is the man behind the machine who is more important and dictates the mission. And this man must be a ground soldier who first learns his military skills and then keeps himself fully abreast with the changing tactics of the ground forces. He is as much part of the ground forces as the Forward Observation Officer (FOO) or the Squadron Commander of a tank regiment or a Company Commander in immediate contact with the enemy, and can think, appreciate and 'read' the battle in the same manner as they do. His flying skill only enables him to fly and position himself tactically, evade the hostile fire both from the ground and the air and exploit the characteristics of the machine optimally.

POSTSCRIPT

Thus, between the two wars, Armies in those countries pushed their cases to get back the control of light aircraft which was strongly resisted by the new independent Service. However, after years of debate and trials during Army manoeuvres and exercises in the UK and USA, just on the outbreak of World War II, the principle that gunners themselves should control their observation and direction of artillery fire from the air was fully accepted and implemented in the US, UK and French Armies and the 'unarmed air observer' came to be known as the Air Observation Post (Air OP) in the British Army and as Liaison Pilots in the US Army. Now flying better machines, they would respond to the call of Ground Commanders at once, become their eyes and ears, passing vital information, directing concentrations of up-to 500 guns, creating havoc on the battlefield and yet surviving even in the most intense hostile air and ground environments. Many a time, they responded and accomplished tasks "beyond the call of duty" with success. As seen from the "after action reports" their contribution was acknowledged and lauded by the Ground Commanders, including the adversaries.

Based on the experience of World War II, the above concept was extended to some other roles as more tools of war were now available which resulted in the formation of the Army Air Corps in all the modern Armies of the world. Although over the years, new technologies and new weapon systems to locate, identify and engage enemy targets have further revolutionised the art of warfare, reconnaissance and observation in the immediate vicinity of the battle area will remain a major concern of the ground commanders in any future conventional war where unarmed air observers, equipped with better aerial platforms will continue to play an important role.

This role performed by the erstwhile Air OP units is now the responsibility of the Reconnaissance and Observation units of the Army Aviation Corps in the Indian Army. The Army aviators of this youngest corps have continued to perform this role with professional skills as done by their predecessors and as visualised by their pioneers, and will continue to do so in the future.

REFERENCES

Unarmed into Battle – The Story of the Air Observation Post by Major General H J Parham CB, CBE, DSO and EMG Belfield, MA.

Soldiers in the Air by Brigadier Peter Mead.

The Eye in the Air by Brigadier Peter Mead.

The Army in the Air – The History of the Army Air Corps by General Sir Anthony Ferrar – Hockley.

United States Army Aviation Digests – June 1962 to May 1963 issues.

Soldiers in the sky Published by Additional Directorate General Army Aviation.

A Few Good Men (History of The Army Aviation Corps) by Lt Gen Anjan Mukherjee and Maj Gen KN Mirji, VSM published by Additional Directorate General Army Aviation.

Notes on the History of Army Aviation by Late Brig R C Butalia.

Archive files of the author.

www.ingramcontent.com/pod-product-compliance
Lightning Source LLC
Chambersburg PA
CBHW021357300426
44114CB00012B/1270